Light Airplane Navigation Essentials

Other Books in the Practical Flying Series

Handling In-Flight Emergencies *by Jerry A. Eichenberger*
Cockpit Resource Management: The Private Pilot's Guide *by Thomas P. Turner*
The Pilot's Guide to Weather Reports, Forecasts, and Flight Planning, 2nd Edition *by Terry T. Lankford*
Weather Patterns and Phenomena: A Pilot's Guide *by Thomas P. Turner*
Cross-Country Flying *by Jerry A. Eichenberger*
Avoiding Mid-Air Collisions *by Shari Stamford Krause, Ph.D.*
Flying in Adverse Conditions *by R. Randall Padfield*
Mastering Instrument Flying, 2nd Edition *by Henry Sollman with Sherwood Harris*
Pilot's Avionics Survival Guide *by Edward R. Maher*
The Pilot's Air Traffic Control Handbook, 2nd Edition *by Paul E. Illman*
Advanced Aircraft Systems *by David Lombardo*
The Pilot's Radio Communications Handbook, 4th Edition *by Paul E. Illman*
Night Flying *by Richard F. Haines and Courtney L. Flatau*
Bush Flying *by Steven Levi and Jim O'Meara*
Understanding Aeronautical Charts, 2nd Edition *by Terry T. Lankford*
Aviator's Guide to Navigation, 3rd Edition *by Donald J. Clausing*
Learning to Fly Helicopters *by R. Randall Padfield*
ABC's of Safe Flying, 3rd Edition *by J. R. Williams*
Flying VFR in Marginal Weather, 3rd Edition *by R. Randall Padfield*
The Aviator's Guide to Flight Planning *by Donald J. Clausing*
Better Takeoffs and Landings *by Michael C. Love*
Aviator's Guide to GPS, 2nd Edition *by Bill Clarke*

Light Airplane Navigation Essentials

Paul A. Craig

McGraw-Hill

New York San Francisco Washington, D.C. Auckland Bogotá
Caracas Lisbon London Madrid Mexico City Milan
Montreal New Delhi San Juan Singapore
Sydney Tokyo Toronto

Library of Congress Cataloging-in-Publication Data

Craig, Paul A.
Light airplane navigation essentials / Paul A. Craig.
p. cm.—(Practical flying series)
Includes index.
ISBN 0-07-013454-5 (h).—ISBN 0-07-013456-1 (p)
1. Navigation (Aeronautics) I. Title. II. Series: Tab practical flying series.
TL586.C69 1997
629.132′51—dc21 97-819
CIP

McGraw-Hill

A Division of The ***McGraw-Hill*** *Companies*

5 6 7 8 9 0 DOC/DOC 0 2 1

ISBN 0-07-013456-1 (SC)
ISBN 0-07-013454-5 (HC)

The sponsoring editor for this book was Shelley Chevalier, the editing supervisor was Fred Bernardi, and the production supervisor was Claire Stanley.

Printed and bound by R. R. Donnelley & Sons Company.

This book is printed on recycled, acid-free paper containing a minimum of 50% recycled, de-inked fiber.

This book is dedicated to
a person who is a critical thinking journalist. A person who is tireless for accuracy. A talented person who has a unique ability to see behind a story. A person who can blend the truth and art and design into near perfection. A person who cares about making every project just right. And for a person that for almost 40 years I have called Mom. That person is Anne B. Craig. Thanks, Mom, for helping me put the "English" in!

Contents

Acknowledgments

This book would never have been written if it were not for the desire and inspiration given from many Aerospace students and faculty members at Middle Tennessee State University. Also the members of the Eagle Squadron from Dickson County High School in Dickson, Tennessee.

The book's original idea had enthusiasm injected into it by Shelley Chevalier, the Aviation Acquisitions Editor at McGraw-Hill. Thanks as always, Shelley.

And of course this effort never would have happened without Anne, Floyd, and Millie at Craig Communications in Nashville, Tennessee. My thanks will never be enough for all you did on this project.

Introduction

Part of what makes flying so much fun is traveling. As a Student Pilot I thought the challenge of flying cross-country to a distant airport was the best part of becoming a pilot. I understood that all those airwork maneuvers, stalls, and all those takeoffs and landings, were very important - but I wanted to get away and see the Earth from above. I wanted to be able to read the ground from the airplane like a driver reads a road map. To meet training requirements, Student Pilots must fly to different airports just to prove that they can. They are not much interested in the city they are flying to, after all, as soon as they get more fuel they turn around and come back. In other words, the flight through the air was more important than the destination. We lose something when the destination becomes more important than the flight itself. Sure airplanes are often used as business tools and the actual flight becomes a necessary evil, but it was not like that in the beginning. From time to time pilots need to get back to their roots. All pilots need to get in the airplane on a Saturday afternoon and fly to an airport where they can eat lunch, talk to the people and enjoy the journey. It does not really matter which direction you fly or which airport you discover, the flight is the fun. Use this book to learn the best part of flying: cross-county flight. See things that non-pilots cannot even dream of. Navigate!

Paul A. Craig
October 1996
Murfreesboro, Tennessee

Light Airplane Navigation Essentials

1
The flat Earth and Eratosthenes

BEFORE WE REALLY GET STARTED IT IS IMPORTANT TO DETERMINE where you are when it comes to understanding the Earth we live on and that we hope to navigate across. Take this simple test to help determine where you stand. Glance at figure 1.1 on the next page before you begin, but do not use the map while you take the test. Once you get started you can only "picture" the Earth in your mind.

1. Compare the land areas of Greenland and South America. Which land area is larger?____

2. Which city is farther west: Los Angeles, California, or Reno, Nevada?

3. If you were to fly from New York City to Chunking, China, which direction is the shortest distance to travel? ____________________________

4. If you traveled due South from Atlanta, Georgia, for 2,000 miles, where would you be? ☐ The Atlantic Ocean ☐ Brazil ☐ the Pacific Ocean

5. While flying the shortest route between Washington, D.C. and Tokyo, what city would you pass over? ☐ Los Angeles ☐ Detroit ☐ Paris ☐ Mexico City

6. When was the Earth first proven to be round?
   ☐ 240 B.C. ☐ 1492 A.D. ☐ 1066 A.D. ☐ 1957 A.D. ☐ 1968 A.D.

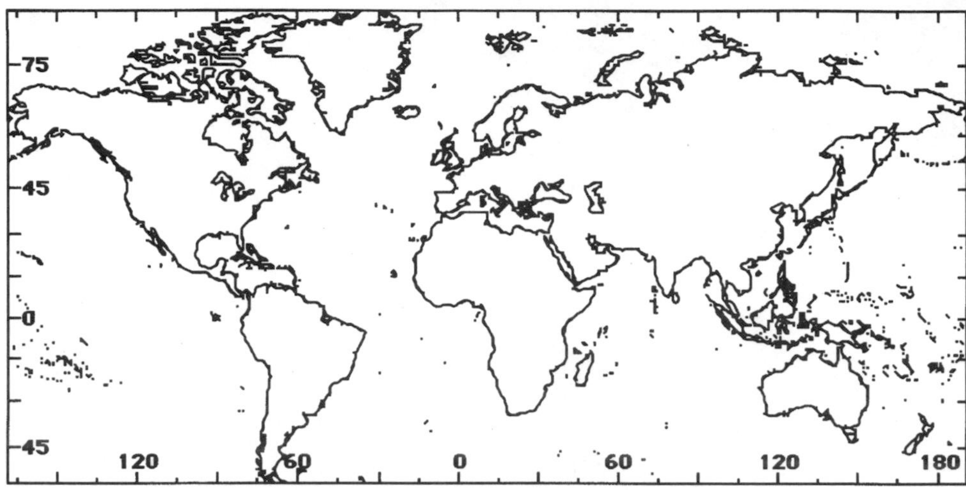

Fig. 1.1

GLOBAL MISCONCEPTION

The map of the Earth seen in figure 1.1 is very familiar. If you walk into just about any grade school or high school classroom in America you can find a map much like this one. It is probably rolled up into a tube just above the chalk board. We all grew up looking at this map. This makes the next statement hard to swallow: This map is wrong! We have been hood-winked all this time. How can a sphere-shape, like the Earth, be printed on a rectangular-shaped, flat map? The answer is that it cannot be with any accuracy. But we have always believed the map to be accurate. We all start with a problem imagining the real Earth. Check and see if your false training on the Earth map has led to some problems on the test.

The answer to question one is: South America. South America is actually about ten times larger than Greenland, but look now at figure 1.1. Greenland actually appears larger. This is because the map is distorted. If you peel an orange and lay out the peel on a flat surface, does the peel form a flat rectangular shape? Of course not. In the same way, the Earth cannot be laid out on a flat piece of paper. The map makers do not want to have gaping holes in their maps however, so what do they do? They lie. They stretch the land masses where gaps exist. Most maps like figure 1.1 are called Mercator Projections, and their scale is only reliable along the Equator. The farther from the Equator the map goes, the greater the distortion. Since the Equator crosses South America, its scale will be more accurate. Since Greenland's location is so far from the Equator, its depiction on the map will be the most inaccurate. But if we believe the map to be true we will have a distorted view of the Earth.

Is Los Angeles or Reno farther west? This one is simple. The answer is...Reno! Even though Reno, Nevada, is inland from the coast and L.A. is on the coast, Reno is still farther west by Longitude.

Which way is shortest from New York to Chunking? When you thought about this question you probably tried to remember if going around the world to China is quickest going to Europe and on east or across to Hawaii and then west. If you chose either east or west then your Earth image is distorted once again. The shortest direction is north, over the pole and then south to China. On a flat map "over the pole" does not make sense. Using figure 1.1, a trip "up" to the pole on the way is an unnecessary detour. But the Earth is a sphere, not a flat paper, so you cannot believe a flat paper's information.

If you transported yourself from Atlanta, Georgia, due South for 2,000 miles you would need a boat, because you would be in the ocean — the Pacific Ocean! How can Georgia, a state with beaches on the Atlantic, be north of the Pacific? Because South America is not located south of North America. The eastern seaboard of the United States is roughly along the same longitude line as the western coast of South America. We Americans tend to think of things in nice tidy rows, so we naturally think that North and South America are aligned. Americans are also very "self-centered," literally. Figure 1.1, and millions of others like it, was obviously printed in the United States, and because we Americans like to think of ourselves as the center of attention, we have been placed in the center of the world map. This despite the fact that this America-centered positioning slices the Earth's greatest land mass (Europe/Asia) into two parts. Do you think school children in Japan or Germany see the world this way?

You should be getting the idea by now. Detroit is the city that lies between Washington and Tokyo along the shortest route. By traveling a "Great Circle" route, hundreds of miles are cut off the journey.

The last question is usually answered the quickest and with the most confidence. Everybody knows the story and the years. The correct answer is...240 B.C.! But you probably answered this question using the poem: "In 1492, Columbus sailed the ocean blue." If you consider yourself an adult and still believe that old story about Columbus being afraid of "falling off the edge," then you once again have been hood-winked. No serious scientist in Columbus' day believed the Earth was flat. Even Columbus did not buy it. The story is simply not true. So what story is true? When was the Earth proven to be round?

THE ERATOSTHENES DISCOVERY

The true story is a very good one and it did take place in 240 B.C. The story begins in ancient Egypt at the library of Alexandria. The library was where the intellectuals of that day hung out. A man by the name of Eratosthenes (pronounced Air-uh taws thin ese) was the "scholar in charge" of the library.

As the story goes, Eratosthenes was touring southern Egypt and stopped by a water well in Syene, which is Aswan today (figure 1.2). By coincidence, he stopped at high noon on the day of the summer solstice. (Solstice means "sun standing still" and occurs on the longest day of the year when the Sun travels directly over the Tropic of Cancer in the North Hemisphere.) As Eratosthenes looked down into the well to pull up a bucket of water he observed a peculiar thing. The Sun's light went straight down into the well without casting any shadows! Eratosthenes had looked down countless wells in his life but had never seen all the way down in bright sunlight. What was different on this particular day? Eratosthenes set up an experiment to find out why.

The next year at precisely the same time of day on the Summer Solstice he was back in Alexandria. At high noon he observed a tower. If the sunlight came from directly overhead again, the tower, like the well, would cast no shadow. But at noon the tower did cast a shadow and Eratosthenes measured the shadow's length. From this alone Eratosthenes solved the great mystery.

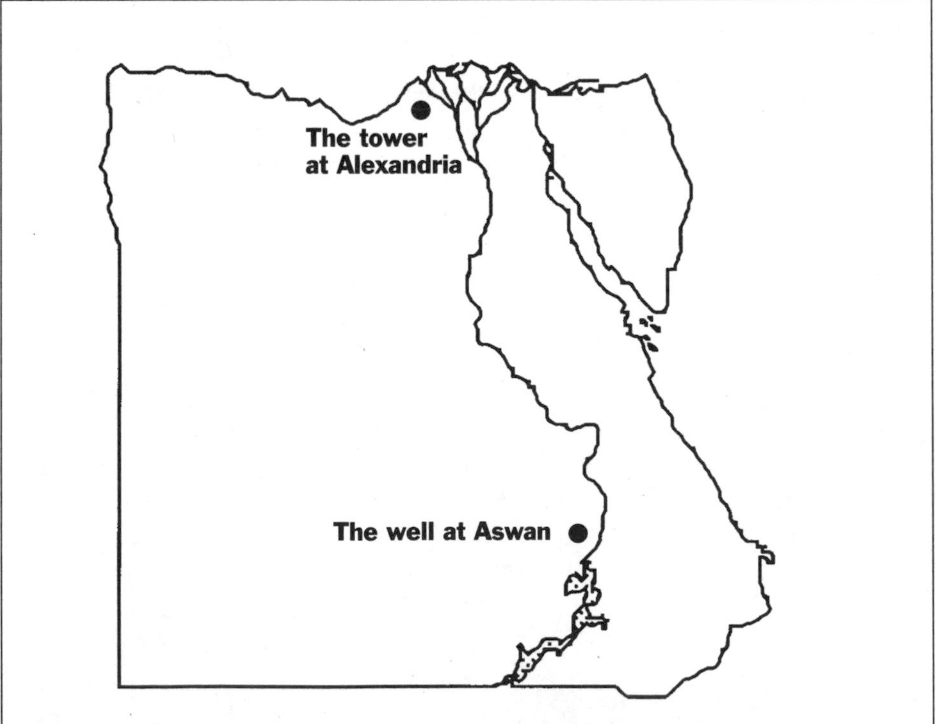

Fig. 1.2

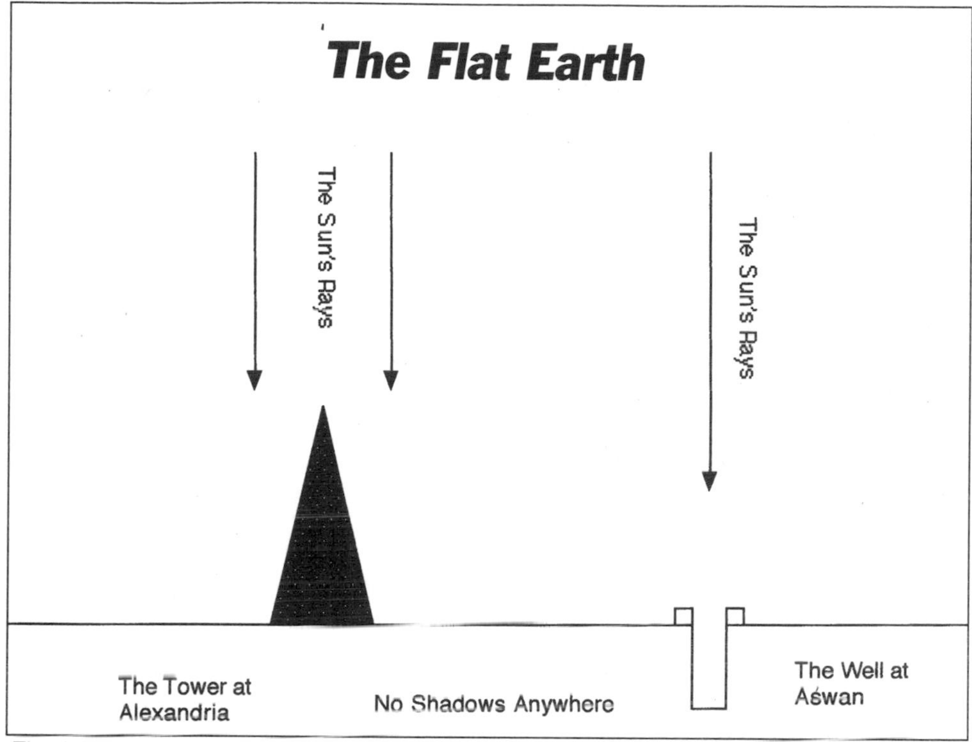

Fig. 1.3

Figure 1.3 depicts the well in Aswan and the tower in Alexandria if the Earth were flat. The result would be no shadow at either location. Figure 1.4 depicts the two locations on the round Earth.

Eratosthenes had proven that the surface of the Earth was curved, but he did not stop here. He measured the angle that the shadow of the tower cast to be 7 degrees, and by using a little geometry he saw that the distance between Aswan and Alexandria was 7 degrees of a complete Earth that would take up a full circle of 360 degrees (figure 1.4). In other words, the distance from Aswan to Alexandria was one 51.4th (360 degrees divided by 7) of the circumference of the Earth.

If the circle of the Earth had been a pizza, (figure 1.5) there would have been 51.4 different slices, each one 7 degrees of the total. To complete the puzzle all Eratosthenes needed was to know the actual distance between Aswan and Alexandria. With the correct distance known he could simply multiply the distance by 51.4 and he would not only know that the Earth was round but also how big the Earth was!

Now in those days there were no scaled maps, so the actual distance was not known. So Eratosthenes hired a guy to walk the trip on a straight line, counting each step! From this crude calculation Eratosthenes calculated the circumference of planet Earth to be 25,000 miles in 240 B.C. The true distance is 24,860 miles. Eratosthenes was a pretty smart guy! Of course Eratosthenes was not experimentally proven correct until Magellan sailed all the way around the Earth, but that happened 1800 years after Eratosthenes. He was definitely ahead of his time!

But if Eratosthenes proved the Earth to be round so long ago, why didn't you already know that? Why were you still believing that "Columbus and the edge"

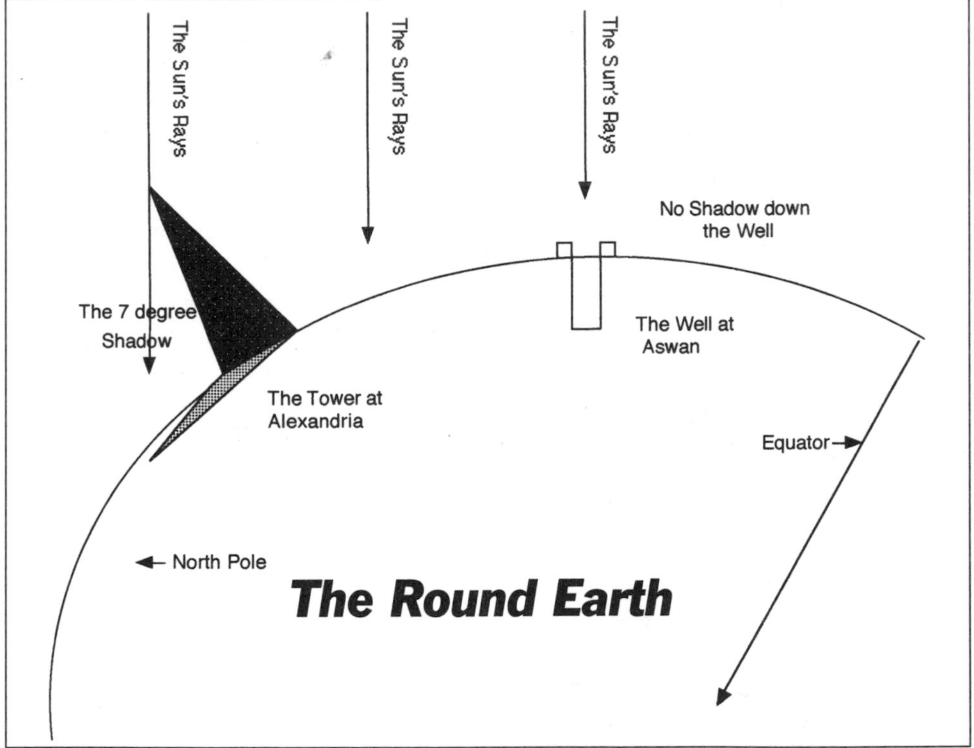

Fig. 1.4

nonsense? The answer may lie in the fact that the work of Eratosthenes was hidden by the emperors of the time. The discovery of Eratosthenes, if true, meant that the world was much larger than previously imagined and much larger than the area of the Roman or Greek empires. If you were the emperor of the world you would not appreciate "Egg-head Eratosthenes" telling you that, in fact, you did not rule the world because there is a whole lot more world out there undiscovered and unconquered. Eratosthenes and his round world were suppressed for political reasons, but now you know the truth.

Why is it important to see the world in a new way: global instead of like a flat map? Because we fly across the curvature of the Earth, not a flat piece of paper.

As pilots, we must think of the Earth as it actually is, not as we have been led to believe. Pilots of the 21st century must think like a man from 240 B.C.

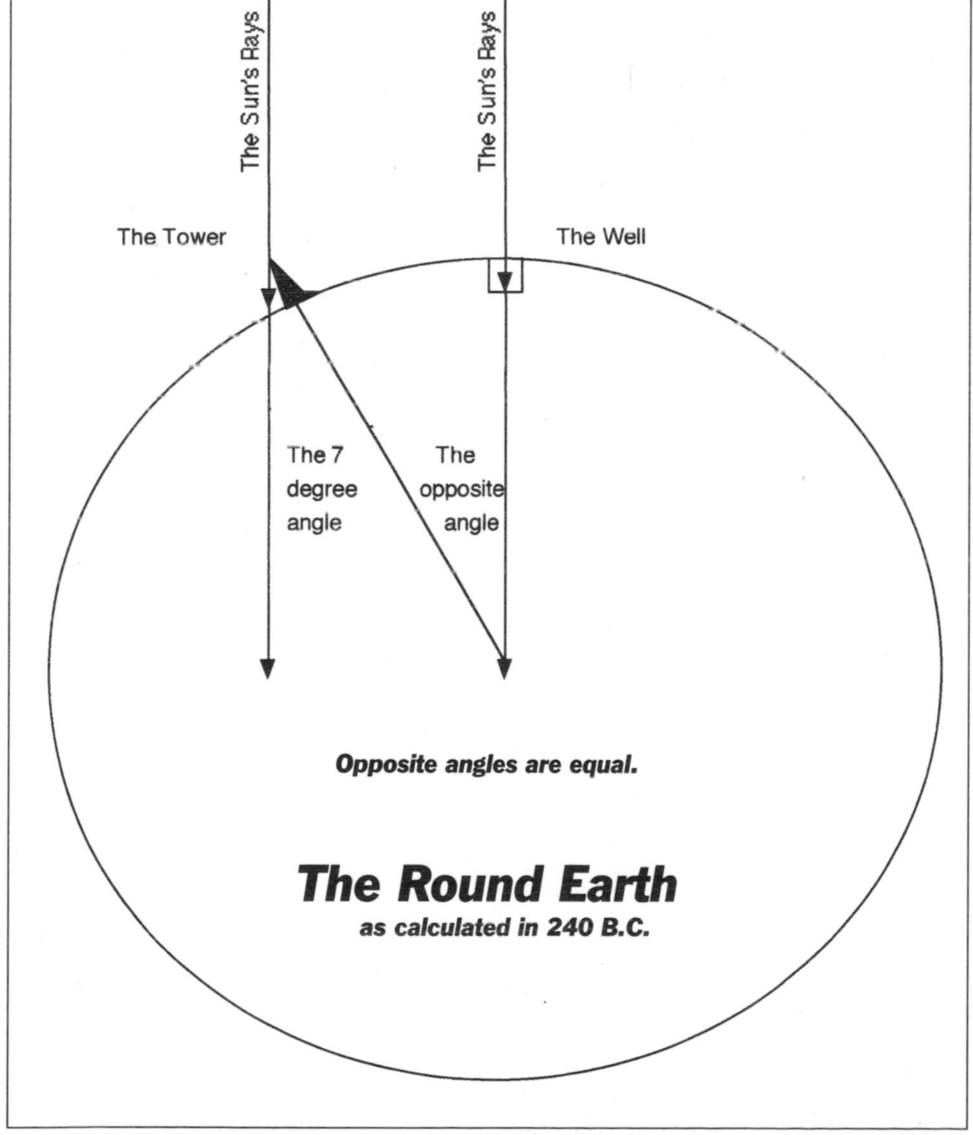

Fig. 1.5

Sec
a

SECTIONAL CHARTS AND SYMBOLS

The Cylinder a
The Mercator

WE HAVE ESTABLISHED THAT THE EARTH IS SPHERICAL, BUT IT WOULD be very hard or impossible for pilots to navigate by holding a globe in their laps. The curvature of the Earth must somehow be represented on paper so pilots can take it with them in the airplane. Figure 2.1 illustrates a cylinder placed down around the Earth. If the Earth's features were then *projected* from the globe and onto the inside of the cylinder you would have a Mercator Projection. As discussed in Chapter 1, this type projection distorts the actual size of the Earth's features except where the globe and cylinder touch: the Equator. Wherever they do not touch there will be distortion. The greatest distortion takes place the farther you travel from the place where they touch: the Poles. This is why Greenland (near the Pole) and South America (on the Equator) look the same size on these charts. Obviously a Mercator Projection is not accurate enough for pilots.

The best solution to the distortion problem is the Lambert Conformal Conic projection. The word *conic* is the key, because this projection is made from a "cone" shape. In figure 2.2 a cone has been placed down over the globe. The inside surface of the cone will touch the surface of the globe. The angles of the cone can be varied so that various surface positions touch the globe. The inside surface of the cone can even be placed *below* the surface of the globe so that one can touch the surface of the globe in two places. These charts, although not perfect, are much better for air navigation. All aeronautical charts use the cone type projection.

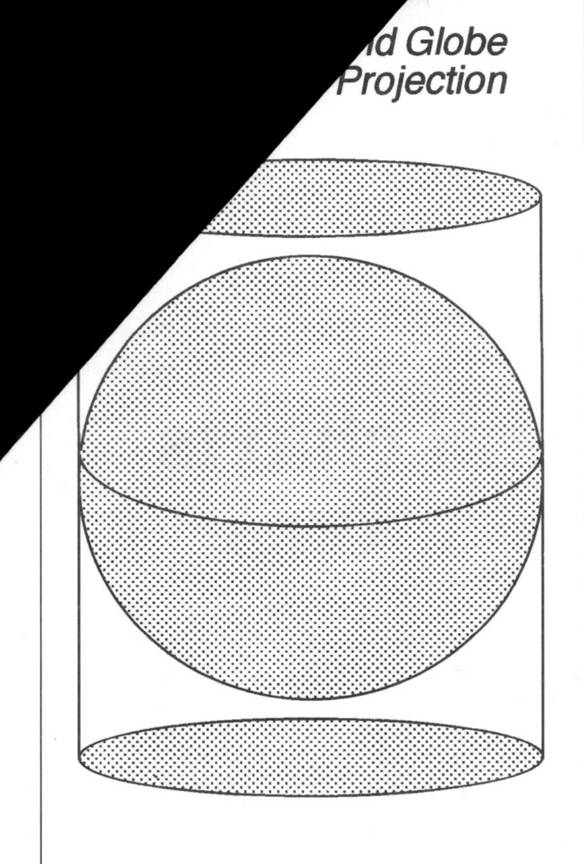

Fig. 2.1

CHARTS FOR NAVIGATION

In the United States the National Ocean Service (NOS) publishes Aeronautical Charts for use by pilots. Several different types are printed for different needs.

The chart that depicts the greatest land area on a single chart is the Aeronautical Planning Chart. These charts have a scale of 1:2,333,232 or 1 inch equals 32 nautical miles. These charts are used to determine mileage for very long flights. Fuel stops can be anticipated, but these charts offer very little detail.

The World Aeronautical Chart (WAC) has a scale of 1:1,000,000 or 1 inch equals 13.7 nautical miles. This chart also is used for longer trip planning. Many individual states print their own aeronautical charts and often these charts are in the WAC scale.

The Sectional Chart has a scale of 1:500,000 or 1 inch equals 6.86 nautical miles and is the type most often used for visual navigation. Sectional Charts (figure 2.3) cover the United States. These charts offer excellent detail of terrain, aeronautical information and airspace.

The chart that has the smallest amount of land area of a single chart is the VFR Terminal Area charts. These charts are printed to depict a particular area of the country that is prone to have a high density of air traffic. The chart's scale is only 1:250,000 or 3.43 nautical miles for every inch of chart space. Today pilots flying VFR in the Los Angeles area are required to have a Los Angeles Terminal Area chart. Although not required at this time, these charts are a very good idea when traveling VFR over other major cities.

The landscape of America is always changing. New roads are built, television towers erected, new airports completed. For this reason no aeronautical chart that is printed today will be accurate for very long. Charts all have an expiration

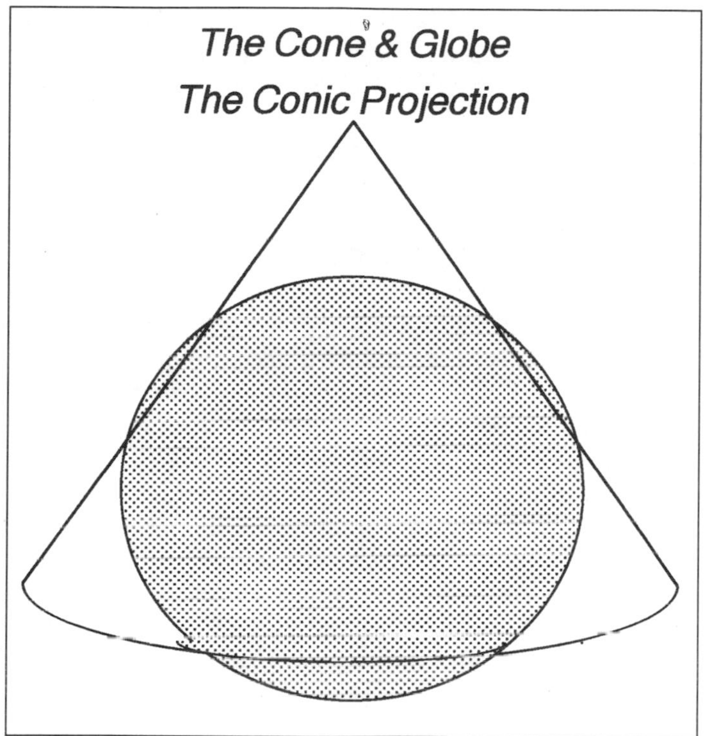
The Cone & Globe
The Conic Projection

Fig. 2.2

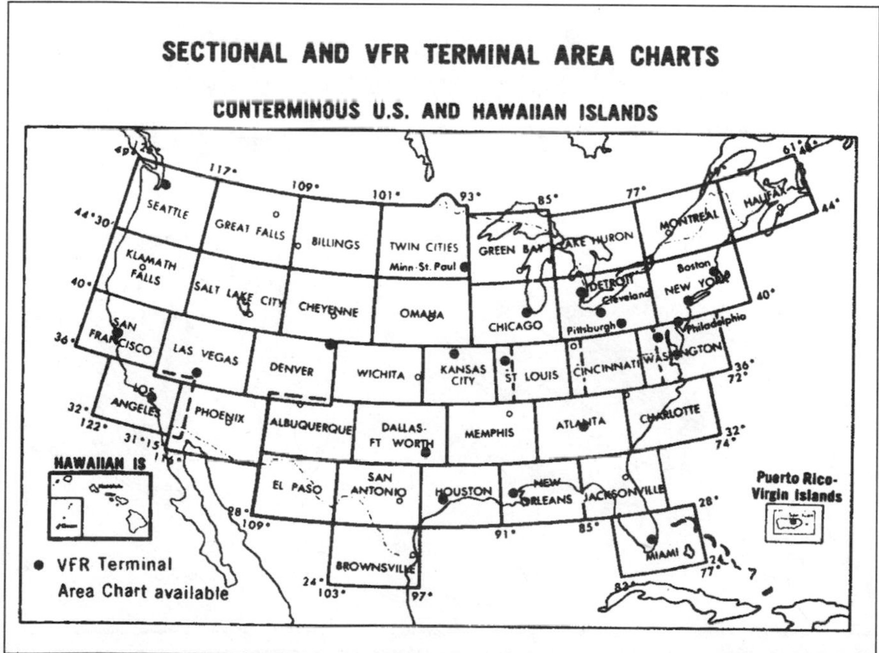
SECTIONAL AND VFR TERMINAL AREA CHARTS
CONTERMINOUS U.S. AND HAWAIIAN ISLANDS
SEATTLE
GREAT FALLS
BILLINGS
TWIN CITIES
Minn-St. Paul
GREEN BAY
LAKE HURON
MONTREAL
HALIFAX
KLAMATH FALLS
SALT LAKE CITY
CHEYENNE
OMAHA
CHICAGO
DETROIT
Cleveland
Pittsburgh
Boston
NEW YORK
Philadelphia
SAN FRANCISCO
LAS VEGAS
DENVER
WICHITA
KANSAS CITY
ST LOUIS
CINCINNATI
WASHINGTON
LOS ANGELES
PHOENIX
ALBUQUERQUE
DALLAS-FT WORTH
MEMPHIS
ATLANTA
CHARLOTTE
EL PASO
SAN ANTONIO
HOUSTON
NEW ORLEANS
JACKSONVILLE
BROWNSVILLE
MIAMI
HAWAIIAN IS
Puerto Rico-Virgin Islands
● VFR Terminal Area Chart available
117°
109°
101°
93°
85°
77°
44°30'
40°
36°
32°
122°
28°
109°
24°
103°
97°
91°
85°
82°
77°
74°
72°
44°
40°
36°
32°
28°

Fig. 2.3

date. Between the day of printing and the expiration date the chart is said to be *current.* Never take an expired chart into the air for the purpose of navigation. When purchasing a chart always look to see if you are getting a current chart.

CHARTS ARE NOT MAPS AND VICE VERSA

There is a big difference between a ***chart*** and a ***map.*** A map only has information and a "flat" layout of cities and roads. Maps are perfectly fine for use on the ground in automobiles, but the pilot of an airplane has more to worry about. A Chart, in addition to information, also depicts *elevation* or *relief.* If a pilot must fly over a mountain to reach the destination, a chart will give information about how high the mountain is. The chart will list the height of the mountain in feet above sea level but it does more. The chart provides *contour lines* and *color coding* to illustrate elevation. Figure 2.4 depicts how this is done. In the top illustration two mountains are shown with a valley in between. The lower diagram depicts a view from directly above (the vantage point of a chart) and contour lines. Each line lists an elevation in feet. If a line has the number 120 next to it this means that 120 feet is the elevation of any point along that line. The closer the lines are together the steeper the rise in terrain elevation.

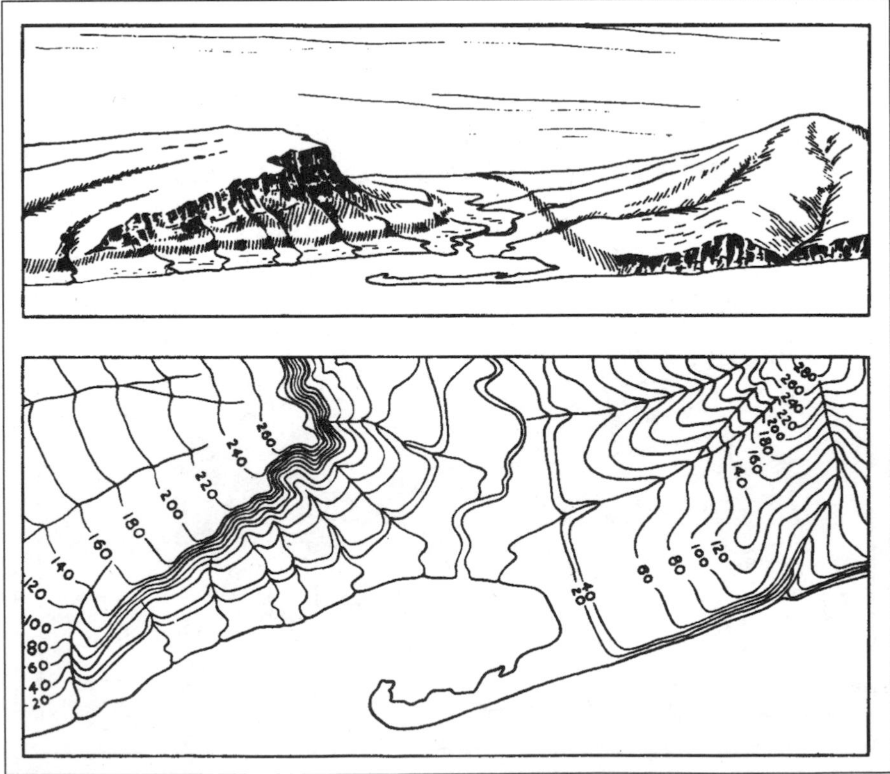

Fig. 2.4

On aeronautical charts the color of the chart changes at every 1,000 foot level. The color of the chart between sea level and 1,000 feet is a pale green. Between 1,000 and 2,000 feet the chart is a darker green. The colors change to yellows, shades of orange, and the highest elevations are shown in brown. The chart colors do not depict the actual colors the pilot will see when flying over! The colors used on any Sectional Chart are shown on the outside cover.

If an aeronautical chart is spread out flat on a table, most people would only see a flat map, but pilots must "see" it differently. By using the contour lines and color coding pilots can easily imagine mountains, valleys, ridges, canyons, and high plains popping off the chart.

AERONAUTICAL INFORMATION

In addition to elevation, aeronautical charts have a huge amount of information and data. This information is divided into categories: Airports and Airport data, Radio navigation aids, Airspace information, Obstructions, Topographical information, and miscellaneous. The charts are not photographs of the Earth. Instead they are only representations. For this reason *symbols* are used to illustrate objects on the ground. Most symbols used on charts are very self explanatory. Tall antennas and towers of the ground have symbols on the chart that actually look like towers. Lakes and rivers are blue, even race tracks are shown as an oval. But some objects on the ground have symbols that require some understanding. A strip mine or rock quarry on the ground is depicted on the chart with a crossed pick and shovel. Forest Rangers use lookout towers to spot fires. These towers are shown on the chart with a triangle inside a circle. Airports are shown as a circle, but as you know airports are usually rectangular to accommodate runways.

Each Sectional Chart has a legend that defines the symbols. Pilots should study these symbols to the point that no hidden meaning escapes them. But some symbols are not completely defined.

What does the *yellow* color on a Sectional Chart depict? Most will answer that this means the presence of a town or city. This is correct but there is more. The shape of the yellow color is supposed to be exactly the same shape as the "light pattern" this city gives off at night as seen from above.

What is the *pennant* defined as *Visual Check Point* in the legend? You might say that this is just a very recognizable location on the ground and could be a very good reference while in flight. This is only partially correct. There are hundreds of very recognizable locations depicted on the chart, but they do not all have a pennant. Any location that depicts the pennant on the chart is also depicted on an Air Traffic Controller's RADAR screen. The location with a pennant will have an underlined name beside it as well. When approaching an area that is served with RADAR pilots should contact air traffic controllers and give their location by referencing the names of these Visual Check Points. The controller is familiar with these locations and when a pilot uses one to announce position the controller knows exactly where to look on the RADAR screen. This will allow the pilot to be RADAR identified faster and therefore safer. Usually pennant locations will ring any airport that has RADAR services.

What about symbols for Gliders, Parachutes, Ultralights, and Hang Gliders? These symbols are shown on the chart in areas where *repeated* operations are known to take place with frequency throughout the year. But these operations can take place just about any place. Pilots should be on the lookout for these hazards every where, not just in the charted locations.

Towers are shown on the chart as if they were laying across the ground. Where is the actual tower location? The actual location of the tower is at the *dot* that is placed at the base of the symbol. Most large towers are held in place by wires that spread out over a wide area. Caution should always be used near towers because the wires are very hard to see.

Why are underground pipelines shown on the chart if they are underground and cannot be seen from the air? The actual pipeline cannot be seen, but the path where the ditch was dug to put the pipeline in can be seen. There will be a long narrow path free of trees.

Sectional Chart Symbols

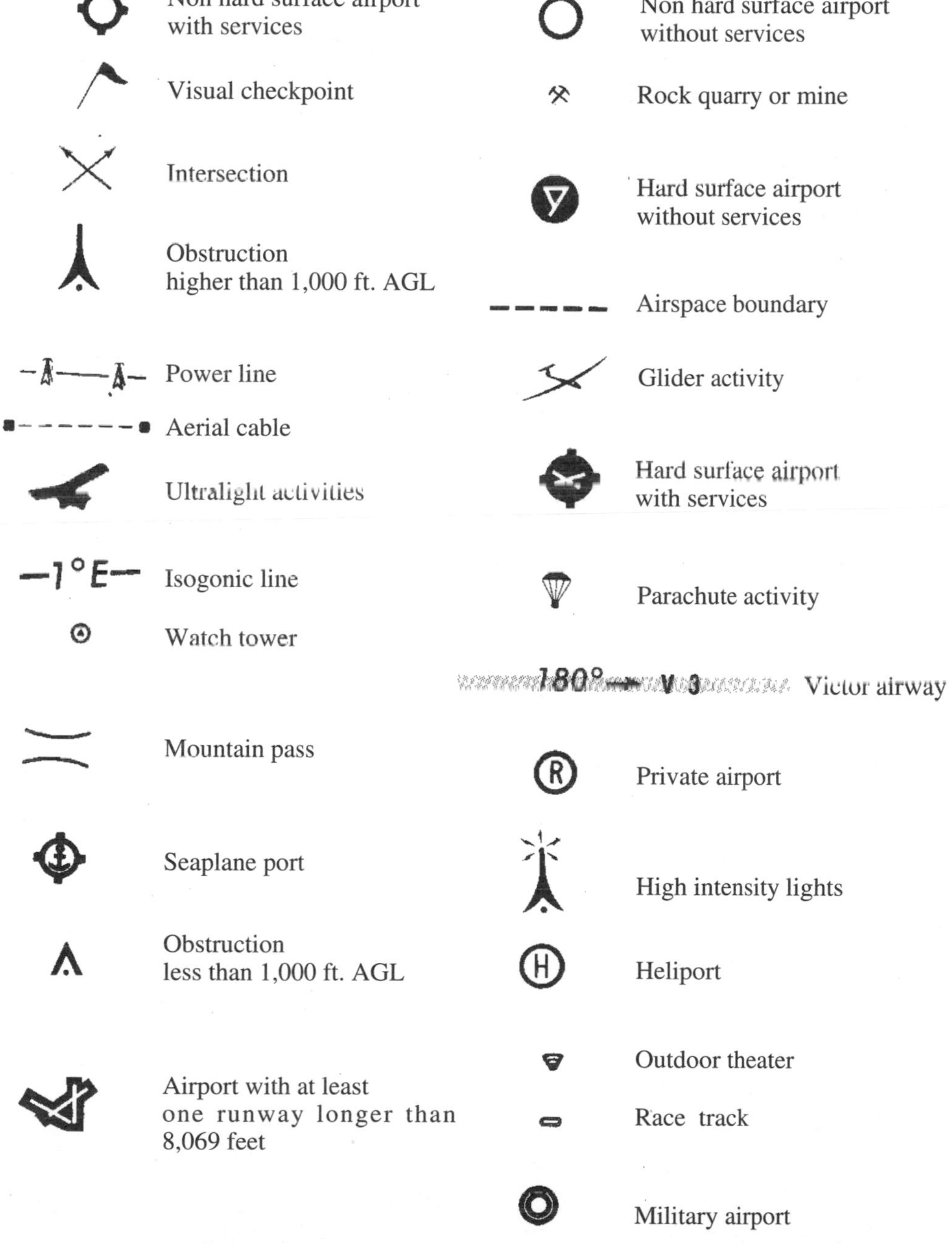

Non hard surface airport with services
Visual checkpoint
Intersection
Obstruction higher than 1,000 ft. AGL
Power line
Aerial cable
Ultralight activities
—1°E—
Isogonic line
Watch tower
Mountain pass
Seaplane port
Obstruction less than 1,000 ft. AGL
Airport with at least one runway longer than 8,069 feet
Non hard surface airport without services
Rock quarry or mine
Hard surface airport without services
Airspace boundary
Glider activity
Hard surface airport with services
Parachute activity
180°
V 3
Victor airway
Private airport
High intensity lights
Heliport
Outdoor theater
Race track
Military airport

3
Latitude and longitude

THE EARTH IS A BIG PLACE AND BEING ABLE TO DETERMINE AN EXACT location on the planet is very important. The system used to determine location on the Earth was developed at the British Royal Observatory. The Observatory opened in 1675 and its astronomers developed the system of Latitude and Longitude. The system lays a grid over the entire Earth. Lines of Latitude run parallel with the Equator and are sometimes called *parallels*. Lines of Longitude run from pole to pole and are also called *Meridians*. Together the lines crisscross the globe and form a grid that can determine exact position. Not surprisingly the designers chose their observatory as the "starting point" of the system called the *Prime Meridian*. The Prime Meridian is the Zero degree Longitude position.

Figure 3.1 illustrates the system of grids. The Equator is 0° Latitude. Traveling north from the Equator the Latitude degrees increase. South Florida is approximately 26° North Latitude, the US/Canada border is about 48° and North Alaska is about 70°. The North Pole is 90° North Latitude.

The Prime Meridian runs through England. The east coast of the United States is approximately 75° West Longitude, Memphis, Tennessee is at 90° West (which means that Memphis is one quarter around the Earth to the west from England) and the International Dateline is at 180° Longitude.

It is not a coincidence that the British Royal Observatory is in Greenwich. The official "starting point" of time is at the Observatory as well. This is where Greenwich Mean Time got its name and why American pilots have the inconvenience of adding or subtracting time on their flight plans. It is a safe bet that if these plans had been made in America the Prime Meridian and Universal Time would have started in Washington, D.C. But remember these systems were invented before the United States was even a country!

THE COORDINATE SYSTEM

Positions on Earth can be determined by pilots by using the Latitude and Longitude grid system. There was a time when this system was not used in everyday VFR navigation, but today the grid is being used more than ever. The Loran and Global Positioning Systems (GPS) that are used now all use the Latitude and Longitude system. The Sectional Charts used for VFR navigation have the grid printed in black on the chart. First look at figure 3.1 of the entire

Fig. 3.1

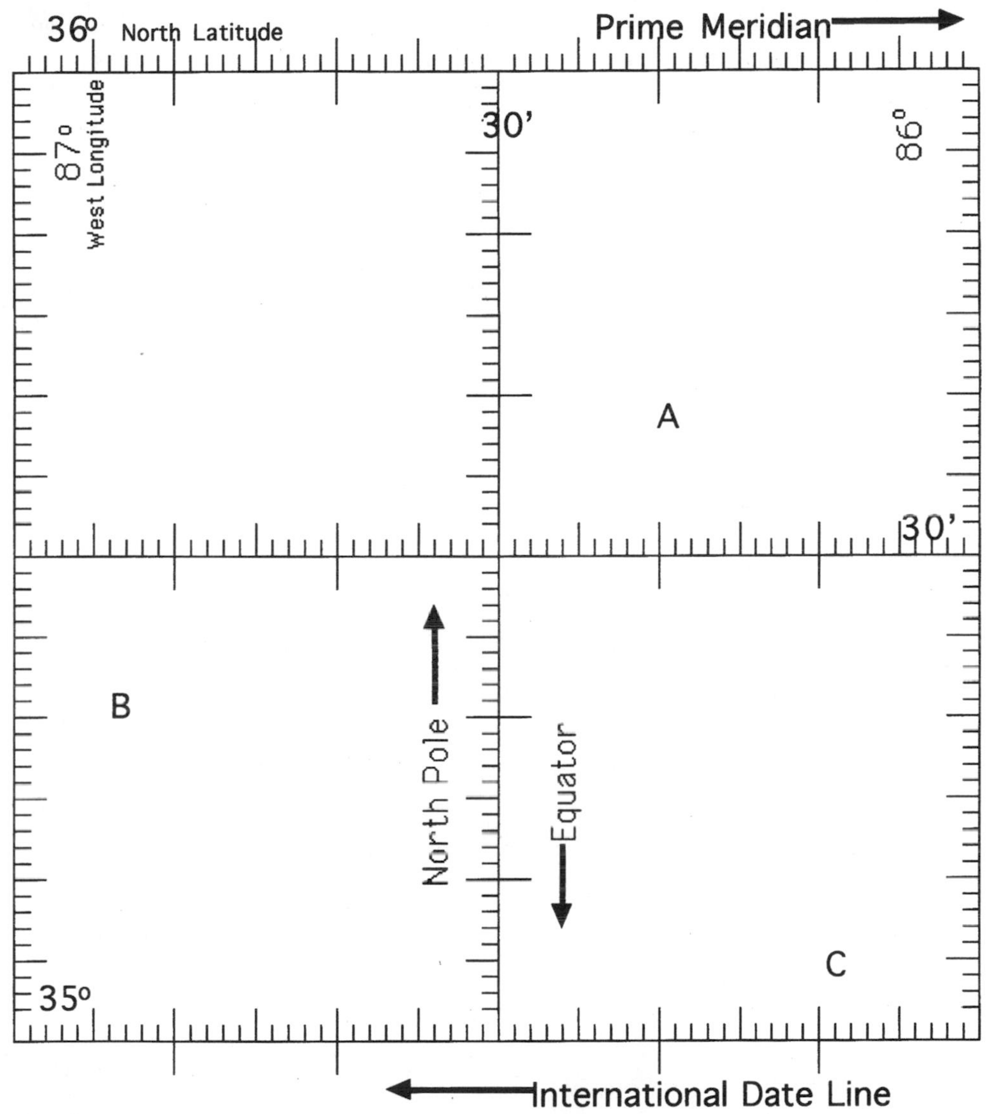

Fig. 3.2

Earth. This global scale is just too large to do VFR navigation, so zero in on just one square. Figure 3.2 is a representation of just one degree of Latitude (between 35° and 36°) and one degree of Longitude (between 86° and 87°).

Each degree is subdivided into smaller units called *minutes*. The term minutes is used, but it does not represent a unit of time in this situation. The term minutes has a symbol as well, an apostrophe ('). When you see the number 30' it is read 30 minutes. There are 60 minutes in every degree. Figure 3.2 indicates the minutes and degrees in a grid just as they are indicated on a Sectional Chart. The actual degree lines are indicated by the number of degrees, like 86°. Halfway between two degree lines is another line marked, 30'. This grid is simply laid over the chart.

Using figure 3.2, determine the Latitude and Longitude coordinates for position A on the diagram. The Latitude numbers are always stated first. Position A is therefore 35° 39' North Latitude by 86° 20' West Longitude. There can only be a single position on Earth with these coordinates. Position A is about 35 and a half degrees north of the Equator and also about 86 and a half degrees west of Greenwich, England. The exact location is 35° 38' north latitude by 86° 20' west longitude. Try your hand at determining the coordinates for positions B and C on figure 3.2.

Now refer to figure 3.3. This is a cut-out from the Atlanta Sectional Chart. The grid from figure 3.2 has now been laid over the chart. Compare figures 3.2 and 3.3 and you will see that they are the same, except 3.3 has all the terrain features, landmarks, and aeronautical information.

Use your understanding of chart symbols from Chapter 2 and your knowledge of the Latitude and Longitude system from Chapter 3 to verify the symbol at the following coordinates. All will be somewhere on figure 3.3.

VERIFY THE SYMBOL AT THESE COORDINATES:

	Coordinates this location	**Symbol discovered at...**
1.	35° 33' North Lat. by 86° 27' West Long.:	Shelbyville airport
2.	35° 27' North Lat. by 86° 47' West Long.:	City of Lewisburg
3.	35° 02' North Lat. by 86° 23' West Long.:	tower 1,994 feet tall
4.	35° 34' North Lat. by 86° 09' West Long.:	church and lookout tower
5.	35° 03' North Lat. by 86° 51' West Long.:	railroad tunnel
6.	35° 43' North Lat. by 86° 16' West Long.:	Cedar Glade private airport
7.	35° 19' North Lat. by 86° 44' West Long.:	mountain with elevation of 1,250 feet
8.	35° 03' North Lat. by 86° 04' West Long.:	visual checkpoint of Fayetteville airport
9.	35° 45.5' North Lat. by 86° 24' West Long.:	rock quarry
10.	Between 35° 00' North Lat. by 86° 41' West Long. and 35° 15' North Lat. by 86° 00' West Long.:	powerline

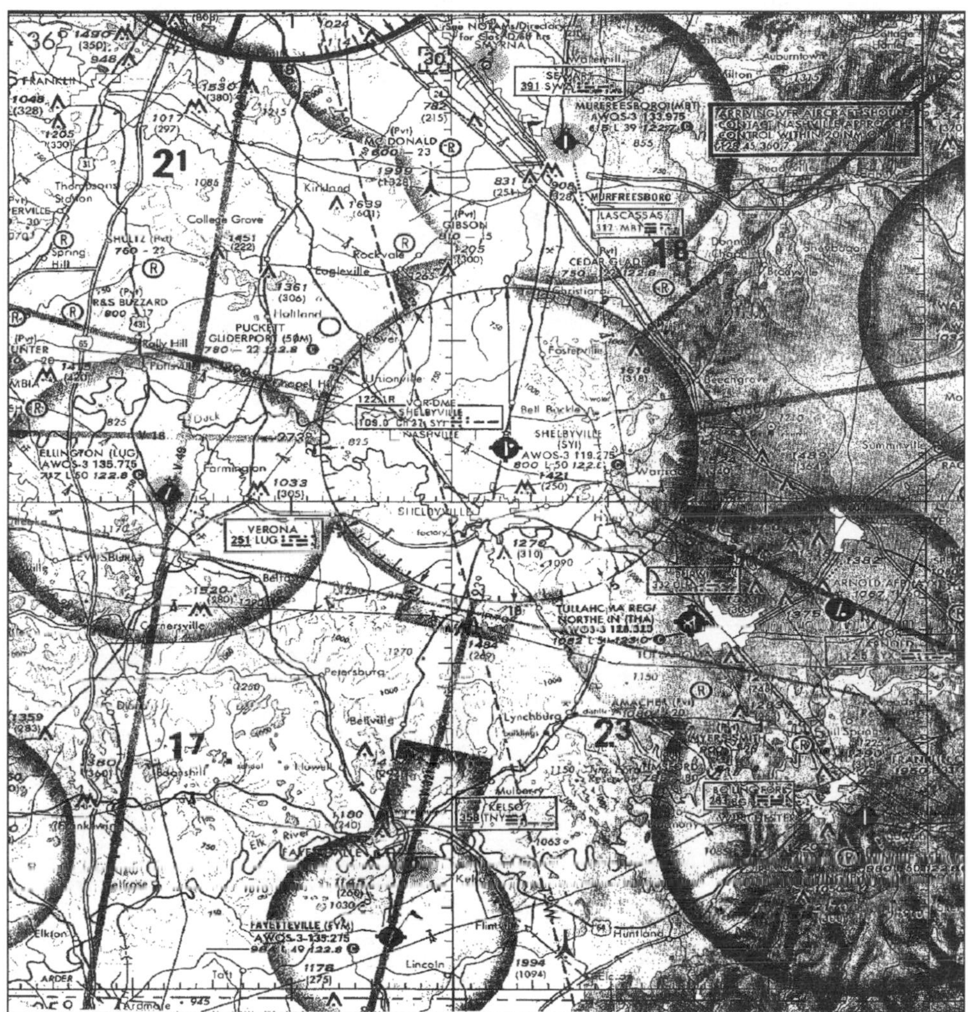
21
17
18
23
College Grove
Kirkland
Eagleville
Rockvale
Unionville
Bell Buckle
SHELBYVILLE
AWOS-3 119.275
Farmington
VERONA
Petersburg
Bellville
Lynchburg
Mulberry
KELSO
Lincoln
Flintville
Huntland
Elkton
FAYETTEVILLE (FYM)
AWOS-3 135.275
MURFREESBORO
PUCKETT GLIDERPORT
ELLINGTON (LUG)
AWOS-3 135.775
Readyville
Auburntown
Summitville

Fig. 3.3

4
The navigation plotter

THE MOST FUNDAMENTAL IDEA OF AIR NAVIGATION IS: WHICH WAY do I go? We are very accustomed here on the ground to following road signs for directions, but in the air there are no signs that point the way. Pilots must rely on compass directions and their ability to aim their airplane in the proper direction. Figure 4.1 is a *compass rose* that shows all 360° of a circle. When flying anywhere the pilot must select one of these directions. The trick is selecting the direction that leads to the destination.

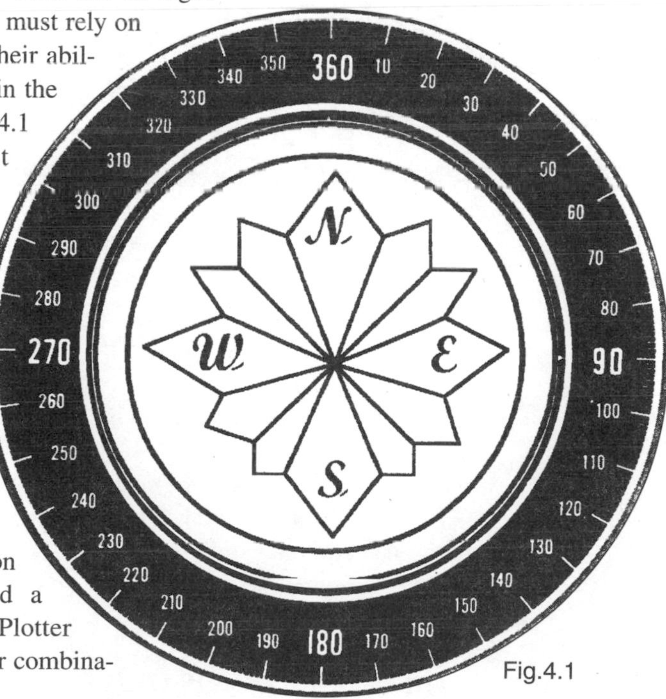

Fig.4.1

To get started, the pilot must use the Latitude and Longitude lines that are on a Sectional Chart and a Navigation Plotter. The Plotter is a plastic ruler/protractor combina-

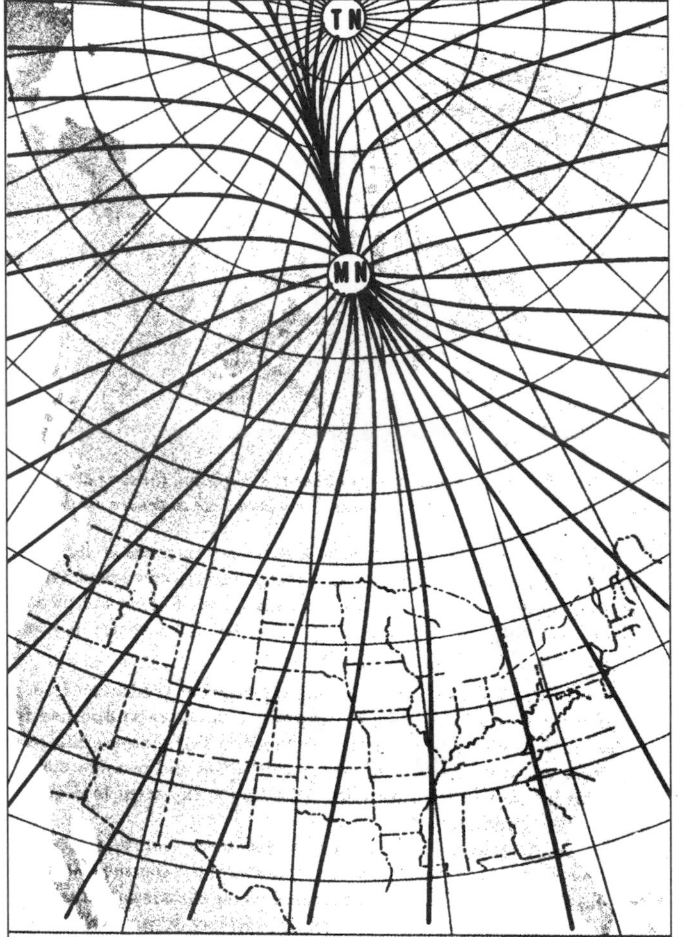

Fig. 4.2

tion. Many manufacturers make plotters and each is somewhat unique, but the basics are the same for any plotter.

The plotter is used to determine a ***True Course (TC).*** This term must be completely understood. Remember back to Chapter 3 on Latitude and Longitude for a moment. If you were to follow any Longitude line to the north, you would eventually arrive at the North Pole. But there is a problem. The Earth has two North Poles! One of these poles is the location where the Earth turns on its axis. If you were to stand exactly on this pole for 24 hours, the Earth would pivot you around. You would face every direction without ever moving your body. This is called the *True North Pole* (figure 4.2). When you use the plotter and a Longitude line to determine direction you will get an answer that is based on True North and is called the True Course.

The second north pole is where the center of magnetism is located. The Earth is like a huge bar magnet with both a north and south pole. This convergence of the Earth's magnetic lines of flux is the *Magnetic North Pole.* Life would be much easier for pilots if these two poles were in the same location, but more on that in another chapter.

USING THE NAVIGATION PLOTTER

Figure 4.3 is a common navigation plotter. The curved half-circle portion is the protractor side. This half-circle is attached to the straightedge part. The straightedge has various mileage scales on all sides. One side of the plotter is usually labeled "Sectional Scale" and the other side "WAC Scale." The pilot must make sure that the correct plotter side is being used with the correct charts. There will be various units of measure for mileage: statute miles, nautical miles, even kilometers. Most all navigation is calculated in nautical miles and should be used unless otherwise requested. When measuring any mileage, first use the correct plotter side, then the correct units, and notice that unlike a ruler the zero miles position is not on the very end of the plotter.

The line between the protractor and the straightedge is called the *base line.* The base line also has a hole in it at the center of the protractor. The base line and hole are used to help determine True Course.

To discover the True Course of a particular flight, first use the Plotter's straightedge like a ruler and draw a line from the departure airport to the destination airport. This is called the *True Course Line.* After the line has been drawn, see if the line cuts across any Longitude lines. If so, place the plotter's hole over the intersection of the True Course Line and any Longitude line. With the hole remaining over the intersection, rotate the plotter so that the base line lies exactly over the True Course Line (figure 4.4). Now hold everything steady. The Longitude line used under the plotter's hole, can be seen through the plastic protractor. The Longitude line will cut through the protractor scale at some point.

Most plotter's protractors have four scales. The two closest to the rim of the protractor go all the way from side to side until meeting the base line. The two scales on the inside of the protractor only go side to side for a short distance. If using a North-South Longitude line as shown in figure 4.4, the outside scales are used. The pilot determines which scale of these to use based on the flight's general direction.

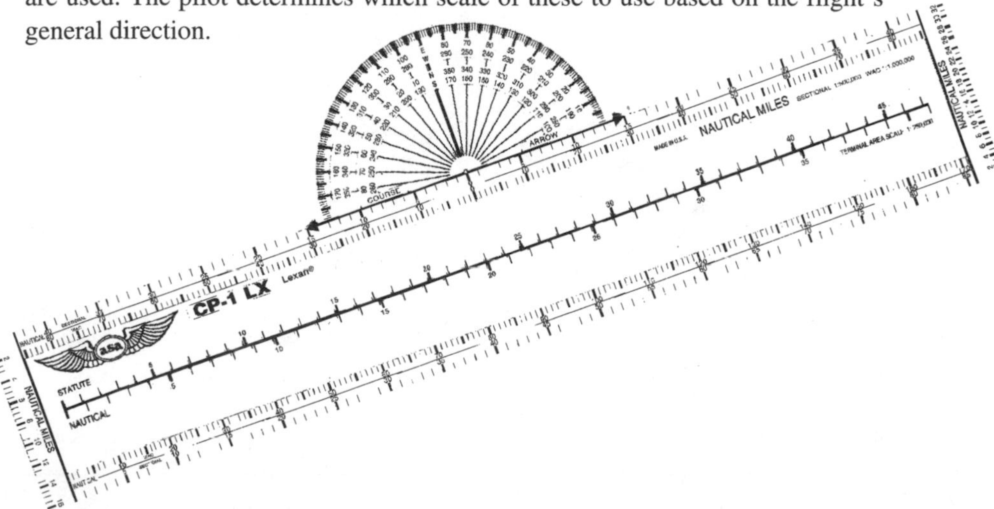

Fig. 4.3 *Courtesy of Aviation Supplies and Academics.*

If the True Course Line does not cross a Longitude line but does cross any Latitude line (figure 4.5) use the same technique as before, placing the hole over the intersection of the True Course Line and any Latitude line, then rotating the plotter to place the Base Line over the True Course Line. Now read the True Course on one of the two inner protractor scales.

It is possible, on short flights, that the True Course Line does not cross either a Latitude or Longitude line. In this case you can still determine the True Course

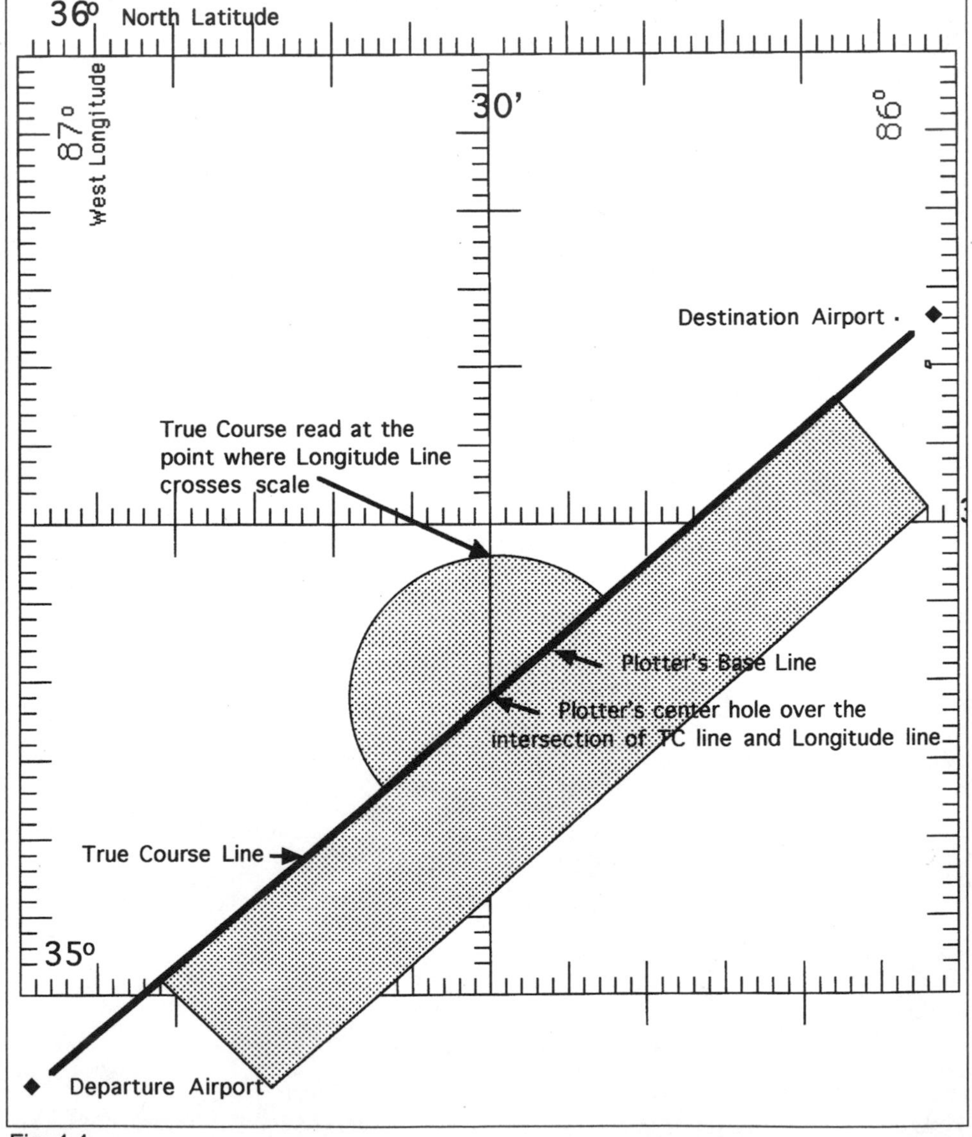

Fig. 4.4

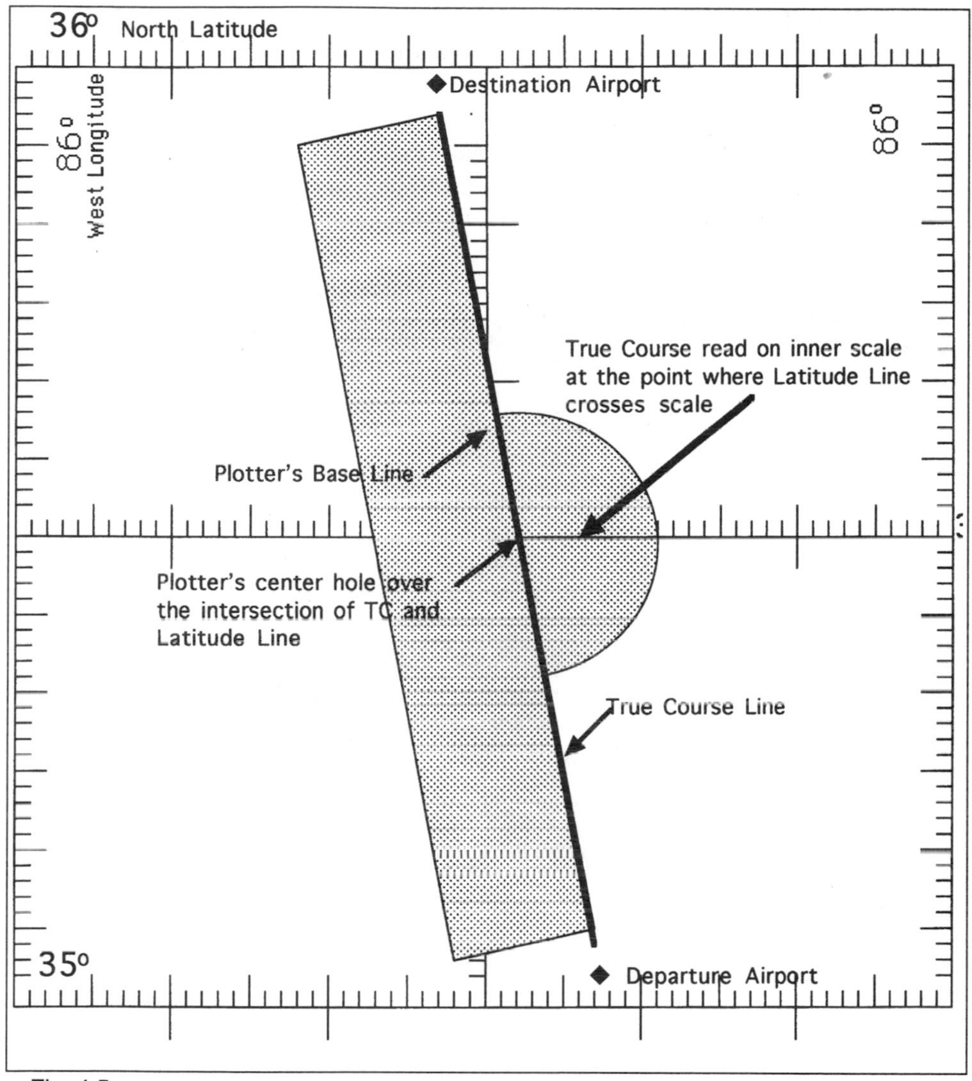

Fig. 4.5

by extending the True Course Line beyond the destination airport until it does cross either a Latitude or Longitude line. From that point plot the course as described above.

PLOTTER PRACTICE:

Refer back to the chart of figure 3.3. Determine the True Course for the following flights.

1. What is the True Course and distance between the Ellington Airport (35° 31' N. Lat. by 86° 48' W. Long.) and the Winchester Airport (35° 11' N. Lat. by 86° 04' W. Long.)? **120°**

2. What is the True Course between the Winchester Airport (35° 11' N. Lat. by 86° 04' W. Long.) and the Ellington Airport (35° 31' N. Lat. by 86° 48' W. Long.)? The return trip! **300°**

3. What is the True Course and distance between the Fayettville Airport (35° 03' N. Lat. by 86° 34' W. Long.) and the Murfreesboro Airport (35° 53' N. Lat. by 86° 23' W. Long.)? **010°**

4. What is the True Course and distance between the Fayettville Airport (35° 03' N. Lat. by 86° 34' W. Long.) and the Looperville Private Airport (35° 46' N. Lat. by 86° 55' W. Long.)? **248°**

5. What is the True Course and distance between the Looperville Private Airport (35° 46' N. Lat. by 86° 55' W. Long.) and the Pucket Gliderport (35° 41' N. Lat. by 86° 38' W. Long.)? **110°**

5

True course and true heading: the computer wind face

THE FACTOR THAT HAS THE GREATEST EFFECT ON THE AIRPLANE IN flight is the wind. There is one basic idea that every pilot must understand and that is: *Airplanes do not always travel in the direction that they are pointed!* The True Course, determined in the previous chapter is simply the direction the pilot points the airplane's nose to the destination. But the True Course will not take you to the destination if there is wind to blow you off. While in flight the airplane will not "weathervane" into the wind, rather it slides sideways even though the True Course might remain constant. Figure 5.1 depicts a pilot who is working hard to hold the True Course, but is not correcting for the wind. While the pilot traveled forward, he also traveled sideways or downwind. The result is that the pilot is off course, may have missed the destination airport altogether, and could soon be lost.

In order for pilots to maintain the proper path to the destination airport, the nose of the airplane must be slightly turned into the wind in order to compensate. This slight turn is called the *Wind Correction Angle (WCA).* The pilot must calculate this angle and apply it to the True Course in order to obtain the proper path. Whenever a Wind Correction Angle changes the True Course it is then called a True Heading. So, TC plus or minus WCA equals TH.

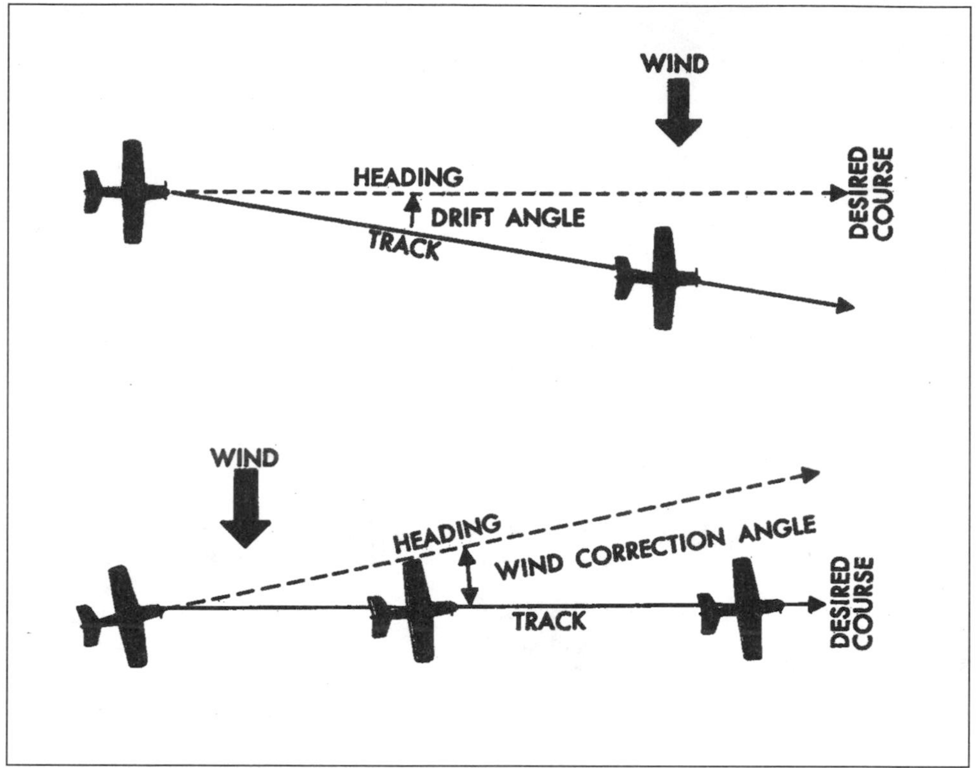

Fig. 5.1

In addition to changing the direction that the pilot points the airplane, the wind can either help or hinder the airplane's progress. If an airplane must fly into a head wind, the plane will go slower across the ground. It is literally like swimming upstream. On the other hand, if an airplane is lucky enough to be flying with a tailwind, its speed across the ground will be accelerated. Every airplane travels through the air at a certain speed. This is called *True Airspeed (TAS).* The calculation of True Airspeed is described in Chapter 7. For now, accept the fact that airplanes move forward through the air, but what if the air itself is moving? Wind effects can increase or decrease the actual speed that the airplane travels across the ground. This is called *Ground Speed (GS).* So, TAS plus a tailwind will make GS faster, and TAS minus a headwind will make GS slower.

WIND DRIFT

So wind affects both direction and speed. If the pilot does not compensate for direction he will eventually be off course. If the pilot does not allow for ground speed variations he might have to stay in the air longer than expected against a headwind and run low on fuel. This idea would be easier if all winds were exact tailwinds, or exact headwinds, or even exact crosswinds, but in the real world the wind can come from any direction. This means the wind offers different *components* of wind effect. The pilot must be able to determine all these components if he or she expects to arrive at the destination on course, on time, and with plenty of fuel to spare. To determine wind effects the pilot can use several methods and tools. There are both manual and electronic flight computers that calculate wind drift, wind correction angle and ground speed. But wind factors were calculated long before these tools were invented. In fact all a pilot needs to determine the wind effect is a protractor, a pencil, and a sheet of paper. The most basic determination of the effect of wind is the Wind Triangle.

THE WIND TRIANGLE

In our high-tech world, filled with gadgets of all descriptions, it may seem hard to believe that a True Heading, a Groundspeed, and a Wind Correction Angle can all be determined with nothing more than a protractor on paper, but it is true. Pilots used this method for decades. Ultimately you may not use a Wind Triangle in your everyday flying, but a look at how a Wind Triangle is constructed will help any pilot understand the basics of wind effect.

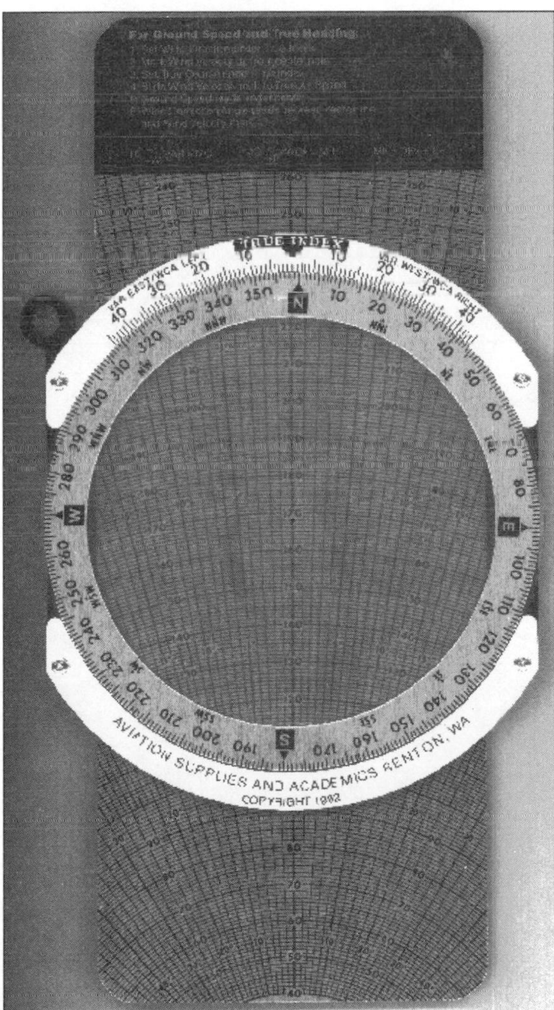

Fig. 5.2 *Courtesy of Aviation Supplies and Academics.*

A Wind Triangle is a visit to your pilot heritage. Start with a set of given information: True Course, True Airspeed, Wind Direction, and Wind Velocity. For example we will use a True Course of 075°, a True Airspeed of 110 knots, and a wind from 025° at 30 knots.

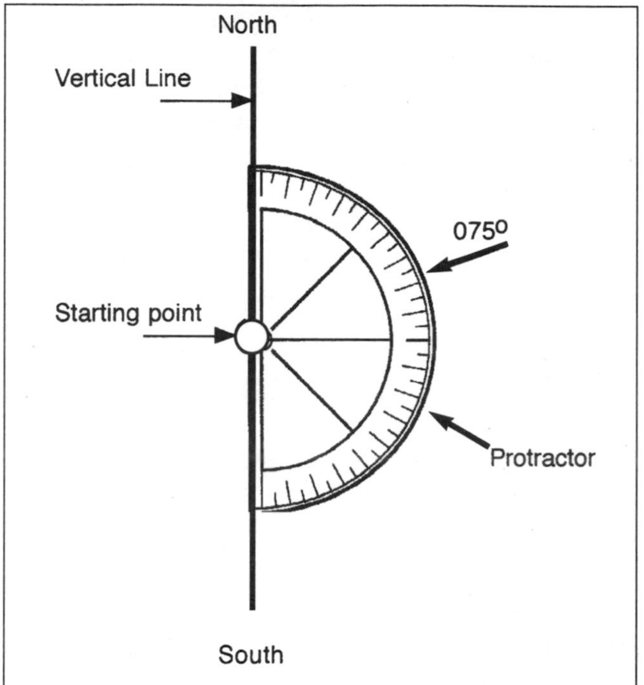

Fig. 5.2a

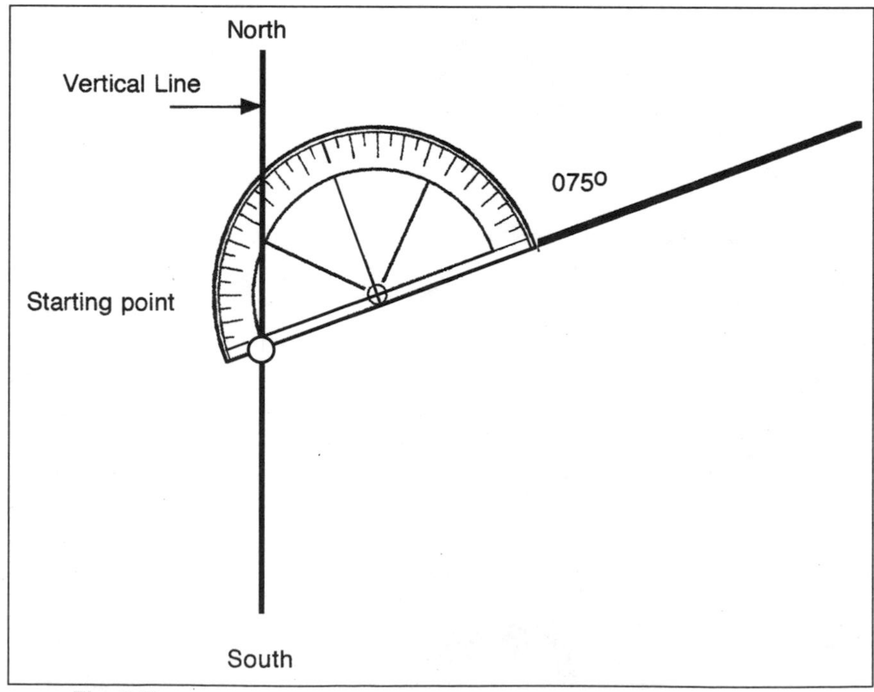

Fig. 5.2b

1. **Start by drawing a vertical line** on the paper to represent the directions of North and South.

2. **Draw the True Course.** Mark a starting point somewhere along the vertical line. Then hold the protractor against the vertical line and mark off 075°. Once the mark is made, use the protractor's straight edge to draw the line out on a course of 075° (figures 5.2A and 5.2B).

3. **Draw in the wind.** The wind "vector" is the force and direction that the wind would carry an object if the object simply floated along with the wind. If a hot air balloon was launched at the starting point in the diagram with a wind coming from 025° at 30 knots, where would that balloon be after one hour? This location is important because this will determine the wind effect. Draw a line that would represent the balloon's path from the starting point with the wind blowing from 025°. Using the protractor mark the 025° point and draw the line *the other direction* to illustrate the direction of wind flow. Now determine a scale for use on the protractor. The scale can be anything as long as the same scale is used throughout the remainder of the problem. One possible scale is one quarter inch (on the protractor straight edge) equals 5 nautical miles. It is also possible to use a Navigation Plotter to construct the Wind Triangle and in that case use the mileage scale on the Plotter's straight edge.

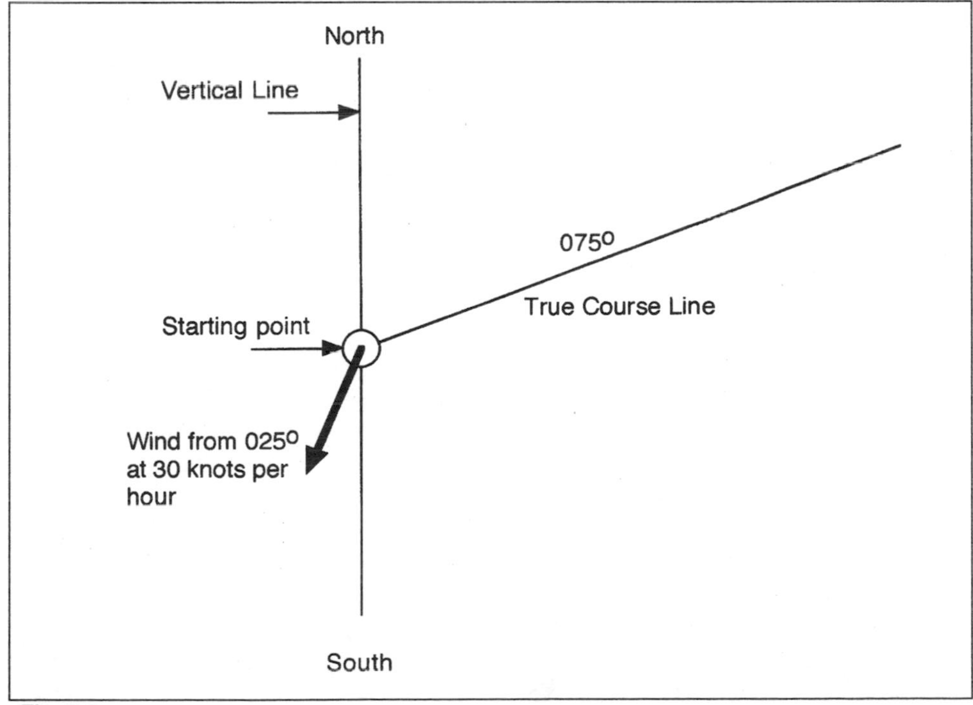

Fig. 5.3

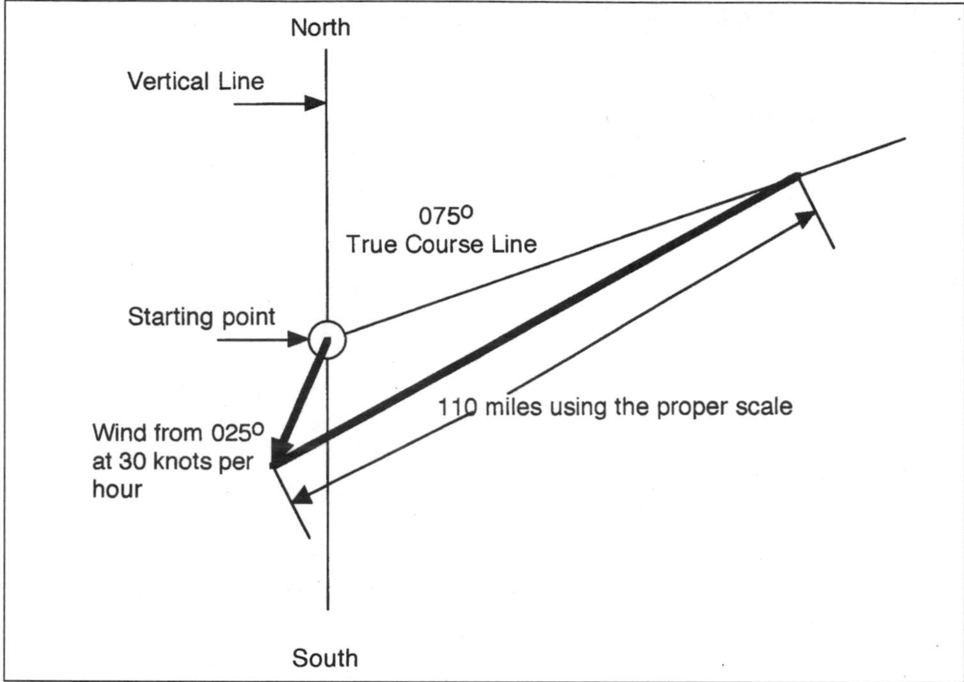

Fig. 5.4

Use the WAC side of the plotter so that the scale will not be too large (figure 5.3).

4. **Complete the triangle.** Using the straight edge of the protractor, or mileage scale of a Navigation Plotter, draw a line that is 110 nautical miles long (using the same scale as in step 3) starting with zero miles at the end of the wind vector and ending where the 110-mile distance crosses the True Course Line (figure 5.4). This line represents the vector or direction and speed of the airplane.

5. **Reading the triangle.** With the Wind Triangle completed the True Heading, Groundspeed, and Wind Correction Angle can all be read from the triangle. The True Heading is the angle between the vertical line and the True Airspeed Line. In this example it is approximately 062°. The Groundspeed is the length of the True Course Line from the starting point to the position where the True Airspeed line crosses the True Course Line. In this example the Groundspeed is approximately 90 knots. The Wind Correction Angle is measured using the protractor and is the angle between the True Airspeed Line and the True Course line. In this example the Wind Correction Angle is approximately 13° (figure 5.5).

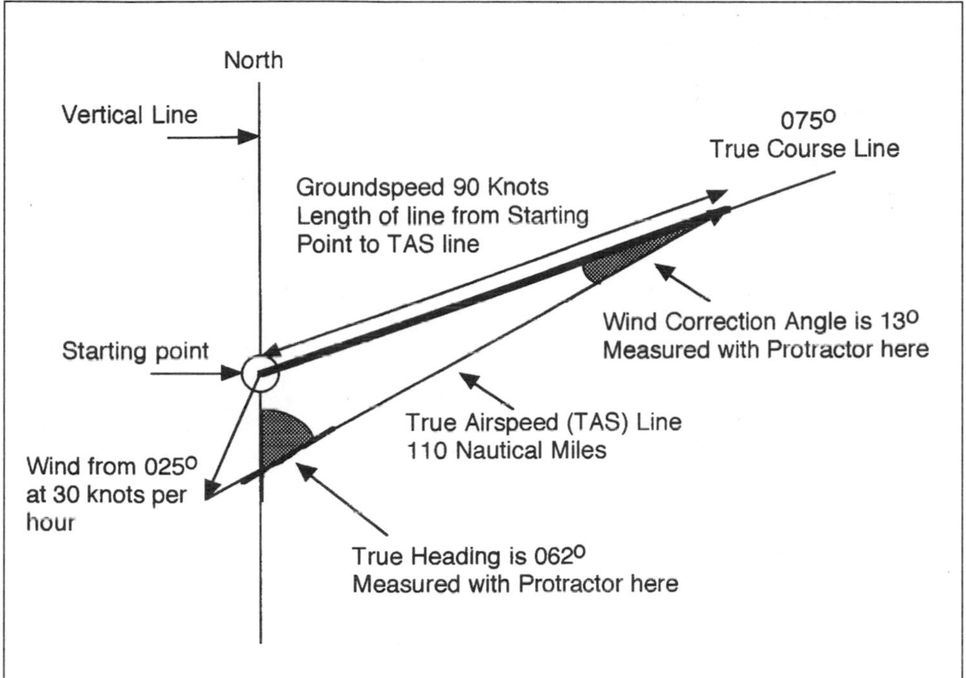

Fig. 5.5

Calculating the groundspeed and fuel burn outbound and then back inbound can lead to the determination of the "Radius of Action." The Radius of Action is the distance an airplane can fly with a given amount of fuel on board, and considering the wind, return to the starting point. In other words how far away from an aircraft carrier can an airplane fly and still make it back to land on the carrier? Using an aircraft carrier in the problem makes the distance determination critical because there is no where else to land. A more practical example would be a "Round Robin" flight. Can a pilot fly from airport A to airport B and return without refueling at airport B? How long can a Search and Rescue airplane stay in a search area before being forced to return for additional fuel? When crossing a body of water, at what location is the "point of no return?" These are additional questions that all spring from the Wind Triangle problem.

Using a manual or calculator type Flight Computer is easier than constructing a Wind Triangle, but understanding the basics of a Wind Triangle leads to a deeper understanding of what you are actually doing when using flight computers. The wind face of the Manual Flight Computer is simply an easier way of constructing a Wind Triangle but it is the same thing.

THE FLIGHT COMPUTER

The Flight Computer can either be a simple cardboard manual wheel and slide card or it can look like a pocket calculator. I personally like the manual computer because the answers received are accurate and it is cheaper. Aviation is already very expensive, so save a little money on your computer. But whether you have the manual or calculator type, make sure you know how to use it. Examples in this book will show the manual flight computer.

The Flight Computer has two sides: the Wind Face side and the Calculator side. The Wind Face side (figure 5.2) has a ring with all 360° of the compass around a transparent disk. Behind the disk is a grid card that can slide up and down behind the transparent disk. Holding the disk in place is a brad or grommet.

Refer to your own Flight Computer for the exact procedure to obtain both the Wind Correction Angle (WCA) and the ground speed. Before you begin the problem you must assemble some data.

You must know :
1. The True Course (TC)
2. The direction the wind is coming from
3. The velocity of the wind
4. The airplane's speed through the air: True Airspeed

Now there is a problem. Only the first entry, True Course, is something we are familiar with. The other three entries have not been fully discussed. The wind's directions and velocity will come from a full pre-flight weather briefing. This briefing is covered in detail in Chapter 14. The last entry, TAS, must be calculated by the pilot. This calculation is covered in Chapter 7. For now we will use "given" information so that you can practice with the wind face. But for real flight planning the pilot will not have any information "given" and must obtain or calculate it all! Once all the data is collected (or in this case "given") the Wind Face can be put to use. Here are the steps to Wind Calculation success!

1. **Place the Wind Direction under the True Index.** Wind is always given by the direction it is coming *from.* In other words if the wind is reported as "180°" or South and you looked directly South, the wind would hit you in the face. The True Index is simply the top of your Wind Face. Rotate the movable disk and place the wind direction at the top, under the mark labeled True Index.

2. **Mark the wind velocity up from the center hole or grommet.** Slide the grid card so that any bold line is positioned exactly behind the center hole. The scale of this card is the same from top to bottom so it does not matter what bold line you use. To make it easy I like to place the bold line labeled

"100" under the center. The lines on the grid card represent velocities. Now, using a pencil, make a dot on the grid card's centerline between the center hole and the True Index. Determine the scale of your grid card. Most cards have a scale where every line is equal to two knots (nautical miles per hour). Count up from the center hole the number of knots that is the wind velocity and make your dot. You can see now why the disk must be transparent, so you can see the velocity lines through the disk.

3. **Rotate the disk and place the True Course under the True Index.** As the disk rotates the wind velocity dot will move to a new location over the grid card. The True Course, that was calculated using the plotter, is rotated to the position under the top or True Index.

4. **Slide the grid card so that the TAS line comes under the wind dot.** The grid card moves up and down like a trombone inside the Wind Face. With a known TAS (Chapter 7), slide the card so that the TAS line on the card rests under the wind dot. When this is done, leave everything alone. Do not move anything. All the answers we are looking for can now be seen.

5. **Read Ground Speed under the center hole.** With everything still in place from after step 4, read what speed happens to be under the center hole. This is the actual speed the airplane will be traveling across the ground. However, be careful! Use a little common sense here. If the airplane has any component of headwind the GS must be slower than TAS. If the airplane has any component of tailwind the GS must be faster than TAS. Double check that the velocity found under the center hole makes sense. Mistakes can be easily corrected by using common sense.

6. **Read the Wind Correction Angle between the centerline and wind dot.** The wind velocity grid card also has lines that are drawn up and down the card. These lines fan out, tighter together at the bottom of the card and farther apart near the top. These lines represent the angles that must be flown to compensate for the wind drift. The wind dot will be resting over one of these lines. The correction angle will be either left or right of the centerline. Use the common sense test here as well. If the wind is coming from the airplane's right side, the wind correction angle should also be to the right.

After completing the Wind Face problem the pilot knows how to compensate for the wind drift. The pilot also will know how fast across the ground the airplane will travel. These numbers are essential to a safe flight.

Work the following problems and verify that your numbers for GS, WCA and TH agree.

	TC	TAS(kts)	Wind Vel(kts)	Wind Dir	GS	WCA	TH
1.	045°	73	10	180°	______	______	______
2.	225°	77	18	345°	______	______	______
3.	135°	83	15	210°	______	______	______
4.	287°	92	22	020°	______	______	______
5.	165°	78	08	175°	______	______	______
6.	250°	80	12	250°	______	______	______
7.	005°	78	25	095°	______	______	______
8.	315°	96	10	135°	______	______	______
9.	110°	84	22	020°	______	______	______
10.	307°	74	18	060°	______	______	______

Answers:

	GS	WCA	TH
1.	80	6R	051
2.	84	12R	237
3.	77	10R	145
4.	90	14R	301
5.	70	01R	166
6.	68	00	250
7.	73	19R	024
8.	106	00	315
9.	81	15L	095
10.	78	14R	321

6

Variation and deviation to compass heading

A *COURSE* BECOMES A *HEADING* WHEN WIND CORRECTION IS APPLIED. The last chapter concluded with the calculation of a *True Heading*. The term *True* refers to the measurement using the True North Pole and *Heading* refers to the fact that a Wind Correction Angle has been applied. With the wind taken into consideration the pilot should be able to navigate to the destination. But wait, there still is a problem! There is no instrument inside our airplanes that tells us which way is *True North*. The only instrument in the airplane that gives information about direction is the Magnetic Compass shown in figure 6.1. (The Gyroscopic Heading Indicator in the airplane does not point to north by itself. The pilot must set this compass by reference to the Magnetic Compass).

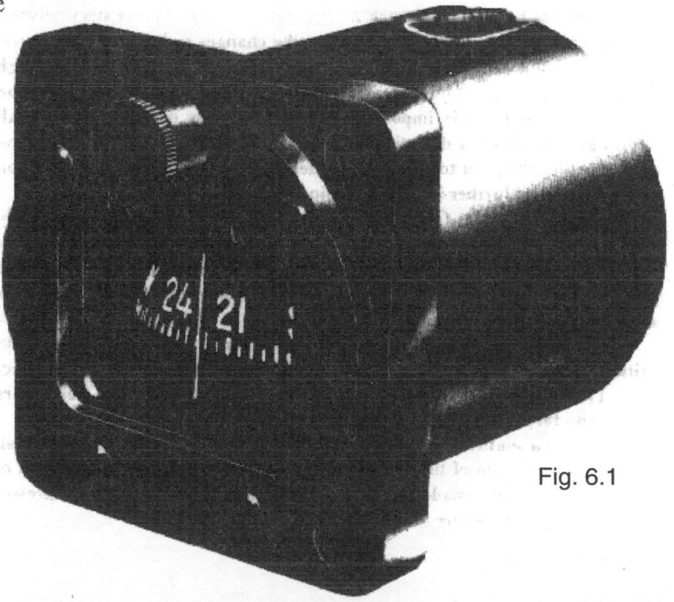

Fig. 6.1

The problem is that the Magnetic Compass points the way to Magnetic North, not True North. As a result the pilot must adjust the heading to allow for this difference in Magnetic and True North Pole positions (refer back to Chapter 4).

VARIATION

Look at figure 6.2. This diagram shows the relative positions of the True North Pole (marked TN) and the Magnetic North Pole (marked MN). The headings determined with a plotter using Latitude or Longitude lines will not coincide with the Magnetic Compass because of this difference in Pole locations. This difference is called ***Variation.*** Imagine that you are flying an airplane at position A in figure 6.2. The direction to True North is completely different than the direction to Magnetic North. The angle of difference is the Variation Angle. If a pilot aimed the airplane's nose toward the True North Pole, the airplane's Magnetic Compass would not be pointing straight ahead. Instead it would be

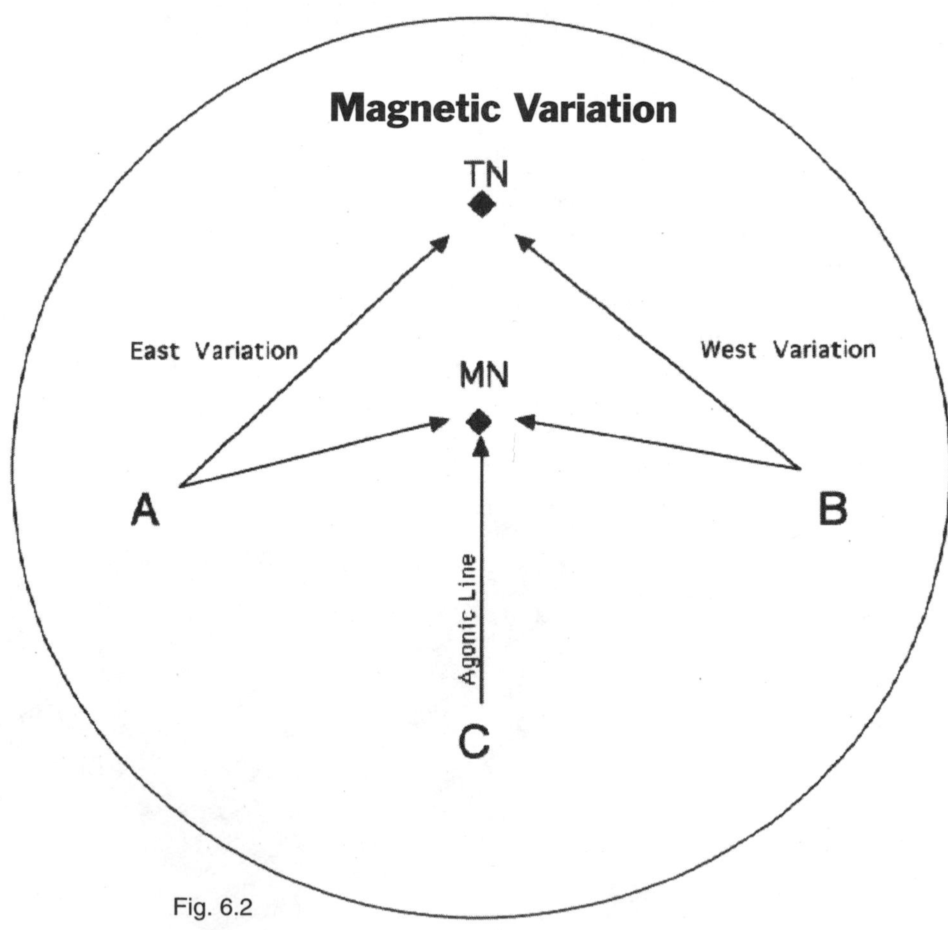

Fig. 6.2

pointing toward the Magnetic North Pole. The Magnetic Compass would actually be pointing more toward the East and therefore this is called East Variation. If you were in an airplane at position B, you would have West Variation. Position C has no Variation because in this position when the Magnetic Compass points to Magnetic North it is also coincidentally pointing to True North.

So that pilots can determine the Variation Angle at their location, lines are drawn on sectional charts called ***Isogonic Lines***. Refer back to figure 3.3. On this Sectional Chart excerpt there is an Isogonic Line. It is shown on the chart as a dash line that runs from North to South across the chart. It will be a magenta (red-purple) color. On figure 3.3 the Isogonic Line is labeled "1°W" along the dashed line. This means that any pilot passing over, or flying along this line must make a 1 degree correction to the West to overcome the Magnetic versus True problem.

The line that has no Variation, or a zero degrees angle, is called the ***Agonic Line.*** The Agonic Line runs through the United States. It runs from North to South on the Eastern side of the Mississippi River.

The Variation Angle should be applied to the True Heading. When this is done the TH changes to a *Magnetic Heading (MH).* But how is this angle applied?

For years pilots have used a simple rhyme to help them remember which way to apply the Variation correction. The rhyme is: *East is least and West is best.*

This means that if the airplane is in a position of East Variation, the angle should be subtracted from True Heading to get Magnetic Heading. If the airplane is in a position of West Variation, the angle should be added to True Heading to get Magnetic Heading. The Variation Angle depends only on location, *not the direction the airplane is heading.* An airplane that is flying in the vicinity of the 10°W Isogonic Line will add 10° to the heading regardless of what the heading is! If the TH is 270° at the 10°W Isogonic Line, then the new MH will be 280°. At the same time, an airplane crossing the 10°W Isogonic Line on a TH of 060° will have an MH of 070°. The Variation does not depend on airplane heading.

DEVIATION

We started by using the plotter and the Latitude or Longitude lines to discover a True Course (TC). Then we added in a correction for wind and changed the True Course to a True Heading (TH). Now we have factored in Variation and by doing so we have changed True Heading to a Magnetic Heading (MH). But we still are not finished! There is one last factor to be considered and that is called ***Deviation.*** Deviation is a change in the direction the pilot must fly due to interference with the airplane's Magnetic Compass. Up until now we assumed that the Magnetic Compass was correctly pointing Magnetic North, but it does not always do that. The heart of the Magnetic Compass is a magnet and magnets can be affected by metal in the airplane or electric currents in the airplane. Look next time you are in the airplane at the position of the Magnetic Compass. It is not placed down inside the instrument panel, rather it is high in the window. This placement is designed to take the Magnetic Compass away from as much interference as possible. The biggest factor affecting the Magnetic Compass is the airplane's radios. The radios have electricity running through them that can create magnetic fields of their own. The created magnetic field is not strong but it is close to the Compass. This can make a compass read incorrectly.

Once a pilot placed a flashlight on the airplane's dashboard. The flashlight had a magnet attached to the side so it would adhere in convenient places. Unfortunately for this pilot, however, the flashlight's magnet severely changed the compass reading, and he flew the wrong way until completely lost. Keep all objects away from the compass!

When the Magnetic Compass is placed into the airplane it is first taken to a Compass Rose which is painted on a taxiway. All radios and electrical equipment are turned on and the airplane's nose aimed at a known magnetic direction. Then the known direction is compared to the reading at the same time on the Magnetic Compass. Any error is recorded. This is done for 12 different directions over the same Compass Rose. This is called "swinging the compass." All the errors are compiled and displayed on a Compass Correction Card. figure 6.3 is a sample card. A card like this will be located near your airplane's Magnetic Compass. The row marked "Steer" indicates what the pilot should actually use to get a Magnetic direction. Figure 6.3 says to "Steer" 117° in order to obtain a Magnetic direction of 120°. This means that the Deviation of

For (Magnetic)	N	30	60	E	120	150
Steer (Compass)	0	28	57	86	117	148
For (Magnetic)	S	210	240	W	300	330
Steer (Compass)	180	212	243	274	303	332

Fig. 6.3

this particular airplane, when facing approximately 120° is a minus 3° (the difference between 120° and 117°).

The planned heading is not always conveniently listed on the Compass Correction Card. What if the Magnetic Heading was 015°? This heading is not directly on the card. In this case you must "interpolate" the card. A reading of 015° is halfway between N (North or 0°) and 030°. The Compass Correction Card of figure 6.3 shows that at North there is a zero degree Deviation error, while at 030° the error is a minus 2° (030° minus 028°). The Deviation error for 015° would be a minus 1° because 015° is halfway between North and 030°, just as a minus 1° is halfway between a zero degree error and a minus 2° error. Sometimes this interpolation calculation will produce a Deviation error like a plus 4.5^{o} or a minus 2.5^{o} error. In these cases it is all right to round up to the nearest degree. Other times the heading in question will not interpolate well, like the previous "halfway" example. In those cases use the *closest* Deviation error.

Unlike Variation, where the airplane's heading does not matter, Deviation does in fact depend on the airplane's heading. And every airplane will have a different Deviation because no two airplanes are exactly alike. Two airplanes that sit side by side on a parking ramp, facing the same direction, will have the same Variation, but they could have very different Deviations.

Now the "direction" calculation has reached a conclusion. TC plus or minus WCA equals TH. TH plus or minus Variation equals MH. And finally MH plus or minus Deviation equals *Compass Heading (CH)*.

Compass Heading is the actual direction that the pilot should fly. CH takes into account all the factors that could change the airplane's direction, or affect it's Magnetic readings.

Verify that your calculations of TH, MH, and CH would agree with these numbers:

	TC +/-	**WCA=**	**TH +/-**	**Variation =**	**MH +/ -**	**Deviation =**	**CH**
1.	085°	7°R	=______	2°E =	______	Figure 6.3 =	_____
2.	240°	10°L	=______	5°W =	______	Figure 6.3 =	_____
3.	022°	14°L	=______	7°W =	______	Figure 6.3 =	_____
4.	318°	17°R	=______	3°E =	______	Figure 6.3 =	_____
5.	005°	8°L	=______	5°E =	______	Figure 6.3 =	_____

Answers:

1. TH = 092°, MH = 090°, Dev from Figure 6.3 = minus 4°, CH = 086°
2. TH = 230°, MH = 235°, Dev from Figure 6.3 = plus 3°, CH = 238°
3. TH = 008°, MH = 015°, Dev from Figure 6.3 = minus 1°, CH = 014°
4. TH = 335°, MH = 332°, Dev from Figure 6.3 = plus 2°, CH = 334°
5. TH = 357°, MH = 352°, Dev from Figure 6.3 = plus 1°, CH = 253°

7

Airplane speeds — IAS to CAS to TAS to GS

THE SPEED THAT AN AIRPLANE TRAVELS IS VERY IMPORTANT IN navigation. First of all, the airplane's speed is it's primary advantage over all other forms of transportation. The speed of the airplane determines how quickly it will reach the destination and this has everything to do with how much fuel will be needed for the flight. If an airplane flies against a strong headwind, it will take longer to reach the destination. This means the airplane's engine will be in operation longer, which means more fuel will be required. The calculation of airplane speed is the first step in making sure that pilots do not run out of gas while still in flight.

You would think that the speed of the airplane is a fixed value, but like a lot of topics in aviation, it is just not that simple. There are actually several "types" of speed that pilots must become familiar with and use. The airplane does not have a speedometer like a car. There is no instrument that reads the airplane's actual speed. Instead, the pilot must rely on a calculation that approximates the exact speed.

AIRSPEED MEASUREMENTS

The airplane does have an Airspeed Indicator, but it has its limitations. First, the instrument produces its information from air pressure. If you stick your hand out the window of a moving car you can feel what is called "ram" air pressure. The faster the car travels the more pressure your hand will feel. The Airspeed Indicator works in much the same way. Ram air is collected by the *Pitot (Pee Tow) Tube* which is mounted on the outside of the airplane and exposed to the on-rushing airflow. The air that comes streaming into the pitot tube is carried through the tubing to the back of the Airspeed Indicator. Inside the indicator there is a diaphragm or expandable balloon. When the air is very fast the balloon is stretched out. When the air is slow the balloon retracts. The needle on the dial of the Airspeed Indicator moves when the balloon stretches.

The problem with using air to determine speed is that the air's density, or number of air molecules in a given area, can vary widely. If the air is thin to begin with, the indicator will not read correctly. The pilot must be ready to compensate for these errors by working through a series of calculations.

The first airplane speed is called ***Indicated Airspeed (IAS)*** and is the easiest to obtain. Indicated Airspeed is simply whatever speed is "indicated" on the dial. The reading may not be correct, and no other factors have been considered, but it is a place to start. While sitting on the ground and beginning the speed calculations you will not be able to read the instrument in flight, yet you must know what the instrument "usually" reads while in cruising flight. You must have some prior knowledge about the airplane you fly. If you do not know or cannot remember what the airplane's Airspeed Indicator usually reads while in cruising flight, ask another pilot or your flight instructor. As for this book, IAS numbers are given, but you must remember where they would actually come from: The Airspeed Indicator.

The Airspeed Indicator and Pitot Tube arrangement has some built-in problems. When the air rushes into the Pitot Tube, it is guided to the indicator through "plumbing." The air must make several turns as it flows to the indicator. The tubing and the amount of corners the air must turn can affect the air's speed. When air is flowing in a tube it will increase velocity when the tube narrows and decrease speed where the tube widens out. Any change in the air's speed through the tube that is different from the actual airflow speed at the Pitot will cause the Airspeed Indicator to read in error. This is referred to as an "installation error." This means that airflows in the tube will predictably be inaccurate. To overcome this problem the instrument's manufacturer makes tests to determine how much error is present at various speeds. The results of these tests are published in the airplane's Pilot Operating Handbook (POH), usually on a table. On this table are the "Airspeed Calibration" errors. When a pilot takes the Indicated Airspeed (IAS) and changes that number to allow for this installation error, the speed number is then known as the ***Calibrated Airspeed (CAS).***

Figure 7.1 is a sample Airspeed Calibration table. The top row is a line of Knot Indicated Airspeed (KIAS). The lower row is the corresponding Knots of Calibrated Airspeed (KCAS). If your airplane usually indicates 100 KIAS while

Fig. 7.1 **Sample Airspeed Calibrations**

KIAS	40	50	60	70	80	90	100	110	120
KCAS	43	51	59	69	77	87	98	108	118

in flight, this table is telling you that the airplane really is not going 100. Instead, due to the installation errors, the airplane is really traveling at 98 knots. In other words, the Airspeed Indicator's plumbing produces a 2-knot error at this speed.

After Calibrated Airspeed has been determined, other factors must be considered. As stated before, the air's density will have an effect on the readings from the Airspeed Indicator. The air density is a measure of the population of molecules that exists in a known area. On a hot day at a high-altitude airport the air molecule density will be sparse. The air is said to be "thin." But on a cold day at an airport on the seacoast (sea level) the air will be very dense and "thick." The thinness or thickness of the air has an effect on the pressure exerted on the Airspeed Indicator. Pilots need to know what the "real" speed is, so pilots must go past CAS and calculate ***True Airspeed (TAS)***. The TAS is the true speed in which the airplane is traveling through the air. To arrive at a number for TAS you first must compare the air's temperature and the air's pressure. These two factors, temperature and pressure, determine the air's density.

The information about temperature and pressure can come from instruments inside the airplane. Temperature information comes from the airplane's Outside Air Temperature Gauge and Pressure information comes from the airplane's Altimeter. Both these instruments can be read accurately when the airplane's engine is not running and all electrical equipment is off.

For calculations the temperature is read off the Celsius scale. The pressure used for calculations is called the *Pressure Altitude* and this is read directly off the Altimeter whenever the number 29.92 is set in the barometric pressure window of the Altimeter.

When these numbers are discovered (from the Temperature Gauge and Altimeter) the True Airspeed can be found by using the Calculator Side of the Flight Computer. The Calculator side has a small window on the circular card. The window reveals an index that is used for Pressure Altitude. On the circular card and just above the window is an index for the Temperature in Celsius. To determine TAS you must first carefully line up the current temperature on the temperature scale across from the Pressure Altitude in the window. Read the scales carefully! When the proper temperature and Pressure Altitude are lined up across from each other, do not turn the wheel. Now look on the outer scale of the wheel for the speed that you previously determined to be the CAS. Across from the CAS on the wheel will be TAS on the outside scale of the fixed card.

Figure 7.2 illustrates the position of the flight computer with a temperature across from pressure altitude and therefore CAS across from TAS. This number is the TAS or the real speed that the airplane is traveling through the air. But what if the air itself is moving? If the air is moving and the airplane is moving in the air, how fast is the airplane traveling across the ground?

Refer back to Chapter 5 for the step-by-step details on how to use the Wind Face side of the Flight Computer to calculate the ***Ground Speed (GS)***. It is the Ground Speed that determines how long the airplane will be in the air and ultimately how much fuel is needed to arrive safely.

The step-by-step procedure to arrive at Ground Speed is diagrammed in figure 7.3. The Indicated Airspeed is obtained in Step 1, by reading the speed shown on the Airspeed Indicator. In preflight planning however, you will not be airborne yet so this speed must come from your memory. You must remember what the airplane indicated while cruising on a previous flight.

The IAS is then taken to the Airspeed Calibration Chart from the Pilot's Operating Handbook (POH) in Step 2. The Calibrated Airspeed is read across the chart in Step 3.

Next calculate the True Airspeed in Step 4 by obtaining the Pressure Altitude from the Altimeter when it is set at 29.92, and the Temperature from the Outside Air Temperature Gauge.

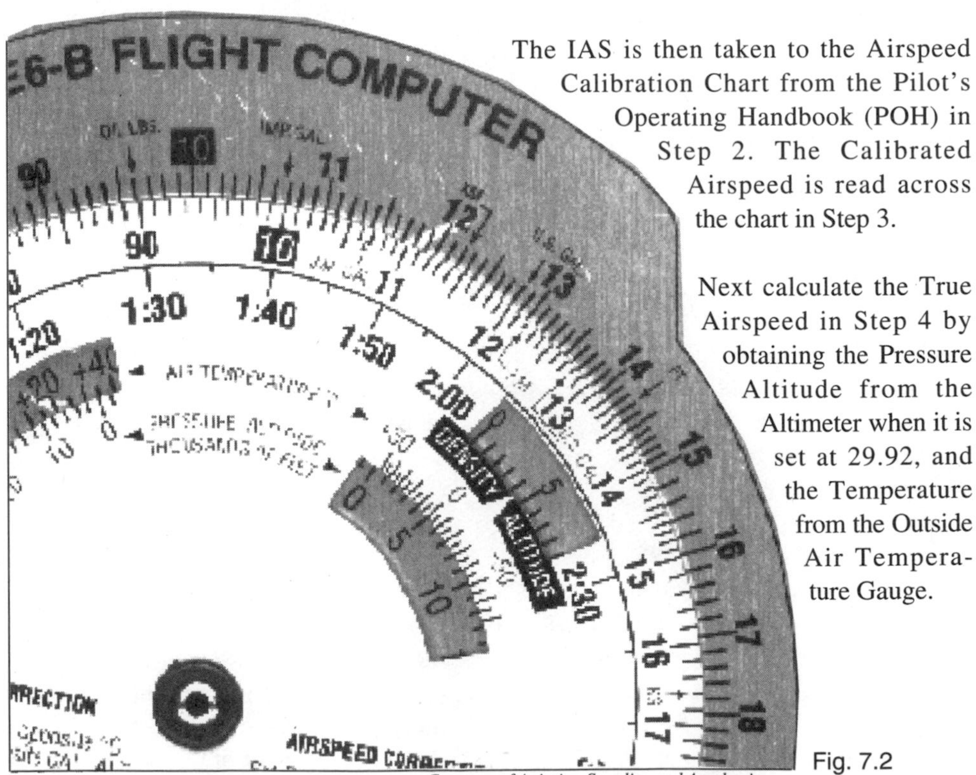

Fig. 7.2

Courtesy of Aviation Supplies and Academics.

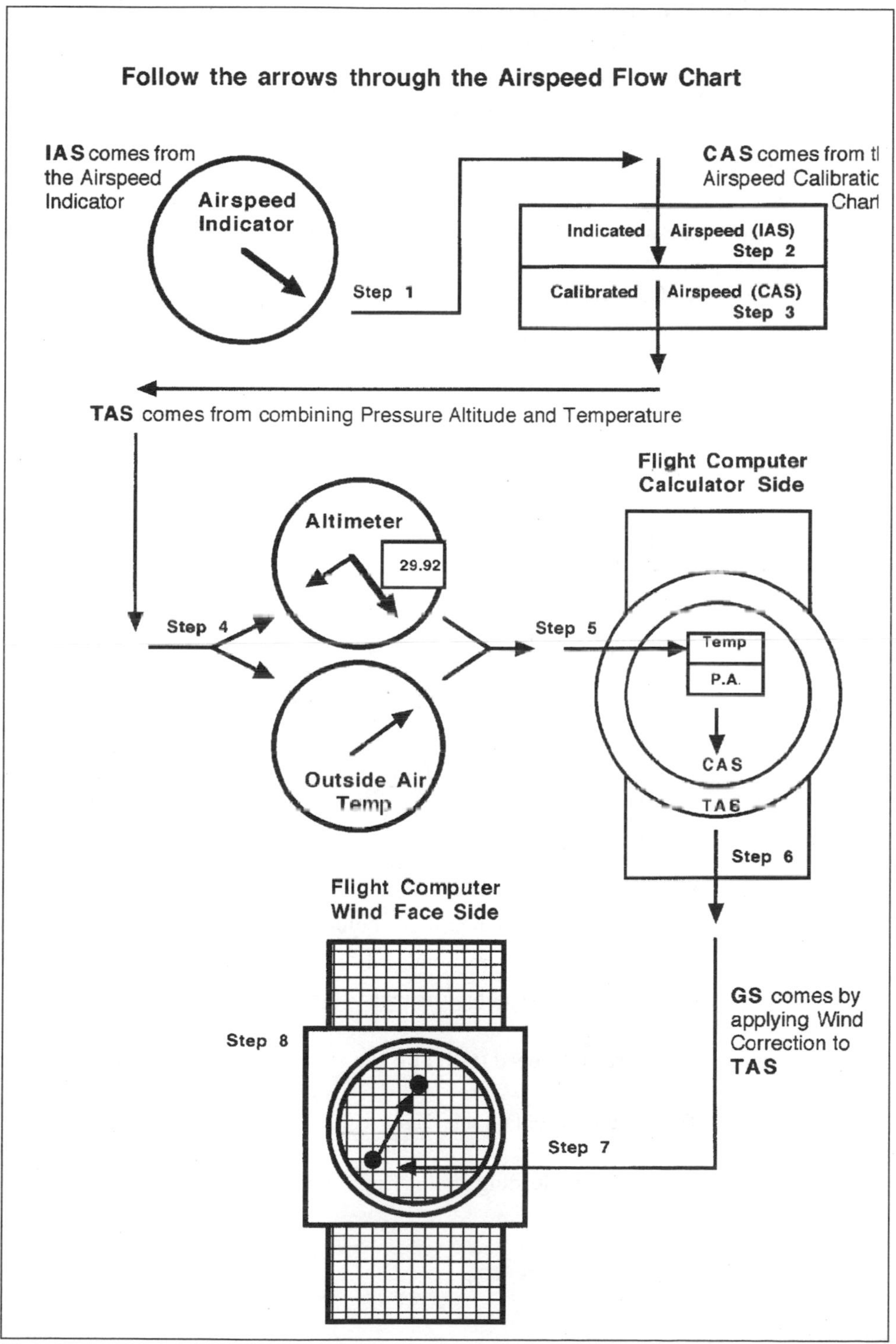
Follow the arrows through the Airspeed Flow Chart
IAS comes from the Airspeed Indicator
Airspeed Indicator
Step 1
CAS comes from tl Airspeed Calibratic Chart
Indicated Airspeed (IAS) Step 2
Calibrated Airspeed (CAS) Step 3
TAS comes from combining Pressure Altitude and Temperature
Altimeter
29.92
Step 4
Outside Air Temp
Step 5
Flight Computer Calculator Side
Temp
P.A.
CAS
TAS
Step 6
Flight Computer Wind Face Side
Step 8
GS comes by applying Wind Correction to TAS
Step 7

Fig. 7.3

Step 5 lines up the Pressure Altitude in the window across from the temperature on the calculator side of the flight computer. When these figures are positioned across from each other, the TAS can be read across from the CAS on the outside of the calculator's scale as seen in Step 6.

The TAS number discovered in Step 6 is then used on the flip side of the flight computer. The "Wind Dot" (Chapter 5) that was placed on the transparent wheel of the wind face is positioned over the TAS in Step 7. While the TAS is in this position the Ground Speed can be read under the computer's center hole. Reading the GS is Step 8.

Reference this Sample Problem using: IAS, CAS, TAS, TC, WCA, TH, Var, Dev, CH and GS.

Start with this information:

IAS read from the Airspeed Indicator = 85 knots

Temperature from the Outside Air Temp Gauge = +68°F

Pressure Altitude from the Altimeter with 29.92 set = 3,000 feet

True Course measured with a plotter on the chart = 294°

Wind Direction = 210°

Wind Velocity = 15 knots

Use the Variation on the Chart - Figure 3.3

Use the Compass Correction Card - Figure 6.3

Use the Airspeed Calibration table - Figure 7.1

Calculate the Ground Speed (GS) and the Compass Heading (CH):

The Ground Speed on this flight will be ___________knots.

The Compass Heading on this flight will be___^{O}.

To Solve the Problem:

To arrive at GS and CH there are a number of items that must be calculated first. Begin with IAS which is given in the problem. Then use the Airspeed Calibration table to get CAS. Then use the Temperature (remember to convert to Celsius) and Pressure Altitude on the Calculator side of the Flight Computer to find TAS. Put the TAS number aside for a moment and go to work on the directions. Use the TC, the wind directions, and the wind velocity, which were given, on the Wind Face side of the Flight Computer to find the Wind Correction Angle and the Ground Speed. This answers the first question about GS, so now finish the calculation to CH. Take the TC and apply the WCA to arrive at the TH. With the TH determined use the Variation to get MH. Last, apply the Compass Correction card Deviation error to the MH to get CH!

Answers:

IAS = 85 knots
CAS = 82 knots
Press Alt = 3,000 feet
Temp = +20°C
TAS = 87 knots
TC = 294°
WCA = 10° Left
GS = 85 knots

TC = 294°
TH = 284°
Var. = 1° West
MH = 285°
Dev = +4° (actually 3.5°)
CH = 289°

8

The navigation record

time/speed/distance, fuel consumption

IN THE PREVIOUS SEVEN CHAPTERS THERE HAVE BEEN QUITE A NUMBER of calculations: IAS, TH, GS, WCA, Var, and all the rest. This is beginning to be a lot of information to deal with. Now is the time to organize this information into a format that can be used in flight. Figure 8.1 is a sample Navigation Record. A Navigation Record is simply a place to record information.

CRM FOR SINGLE-PILOT OPERATIONS

The term Crew Resource Management (CRM) is becoming familiar to aviation. The term refers to a situation where all the crew members on a flight deck work together, drawing all "resources" and the result is a safer flight environment. CRM is thought of as a "big" airplane idea, but it also has applications in smaller, single-pilot airplanes. The pilot of a single-engine airplane preparing to go into the air must also do everything to make the flight environment safer. One way of doing this is to organize your information and your cockpit. The airplane must become a little office where everything is readily accessible. The less time the pilot spends shuffling papers and folding charts while in flight the better. The Navigation Record helps organize the cockpit/office for flight and therefore is a part of CRM.

The Navigation Record of figure 8.1 is designed for efficiency. There are two main sections: "Pre-Flight Calculations and Planning" and "In-Flight Navigation." As the names imply the left side of the Navigation Record would be completed prior to takeoff, while the smaller right side would be completed in the air.

When preparing this Navigation Record the pilot starts on the left side and works to the right. The columns are placed left to right in the same order as the items discussed in the first seven chapters. Columns 2 through 16 of the Navigation Record represent nothing new, only a place to put the information.

Columns 17, 18, and 19 of the Navigation Record do present something new. To navigate properly and with an adequate and legal supply of fuel, the pilot must know how far is the destination? and how much time will it take to get there?

In Chapter 4 the True Course Line was discussed. This line is drawn across the chart from the departure airport to the destination airport. During the pre-flight planning the pilot should look along this True Course Line at the terrain and landmarks that the flight will cross. The line will probably over-fly several recognizable landmarks along the way. These landmarks could be other airports, cities, major highways, rivers, or lakes. Figure 8.2A illustrates a True Course Line from one airport to another with prominent landmarks circled. The landmarks chosen will be called *Checkpoint*s.

When flying cross-country VFR, the pilot must look out the window and spot landmarks along the way to verify that he or she is on course. This "looking out the window" form of navigation is called *Pilotage*. The smart pilot will use Pilotage together with the calculations of Compass Heading to find their way without getting lost. Using courses and compensating for wind drift as this book has done in the first seven chapters is referred to as *Dead Reckoning* (DR) *Navigation*. Pilotage and Dead Reckoning are the two basic forms of navigation that VFR pilots use to get where they are going.

CHECKPOINTS

The Checkpoints are listed in the far left column of the Navigation Record. The first Checkpoint is always the departure airport, because that is where all calculations begin. The next Checkpoint is listed as "Top of the Climb." In Chapter 9 the topic of climb performance is discussed. In that chapter the pilot will learn how to utilize this "TOC" row. After the Top of the Climb Checkpoint comes all the landmarks that the pilot selects along the True Course Line. Figure 8.2B is a Checklist column filled out for the flight in figure 8.2A.

The Checkpoints selected must have good "landmark value." In other words, they must be easily seen and recognized from the air. A rock quarry by a four-lane highway on the northwest side of a city is a good checkpoint. But a creek passing under a powerline is a very poor checkpoint. Try to find good check-

The VFR Navigation Record

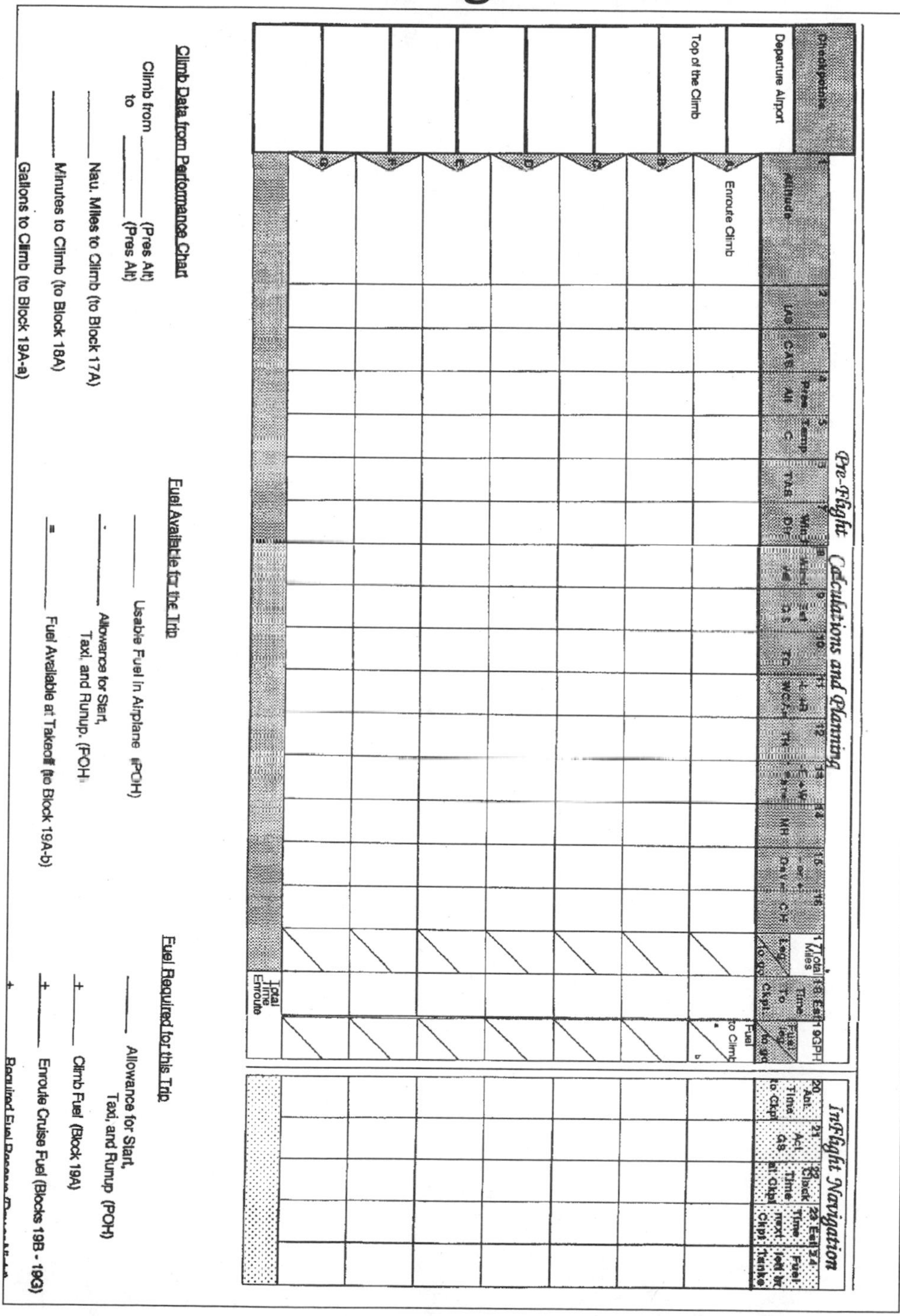

Pre-Flight Calculations and Planning / *InFlight Navigation*

Checkpoints	1 Altitude	2 IAS	3 CAS	4 Pres Alt	5 Temp C	6 TAS	7 Wind Dir	8 Wind Vel	9 Est GS	10 TC	11 +L -R WCA	12 TH	13 -E +W Var	14 MH	15 - or + Dev	16 CH	17 Total Miles Leg / to go	18 Est Time To Ckpt	19 GPH Fuel	20 Act Time to Ckpt	21 Act GS	22 Clock Time at Ckpt	23 Est Time next Ckpt	24 Fuel
Departure Airport	A Enroute Climb																		Fuel to Climb a / b					
Top of the Climb	B																							
	C																							
	D																							
	E																							
	F																							
	G																	Total Time Enroute						

Climb Data from Performance Chart

Climb from ______ (Pres Alt)
to ______ (Pres Alt)

______ Nau. Miles to Climb (to Block 17A)

______ Minutes to Climb (to Block 18A)

______ Gallons to Climb (to Block 19A-a)

Fuel Available for the Trip

______ Usable Fuel in Airplane (POH)

- ______ Allowance for Start, Taxi, and Runup. (POH)

= ______ Fuel Available at Takeoff (to Block 19A-b)

Fuel Required for this Trip

______ Allowance for Start, Taxi, and Runup (POH)

+ ______ Climb Fuel (Block 19A)

+ ______ Enroute Cruise Fuel (Blocks 19B - 19G)

+ Required Fuel Reserve (Day or Night)

Fig. 8.1

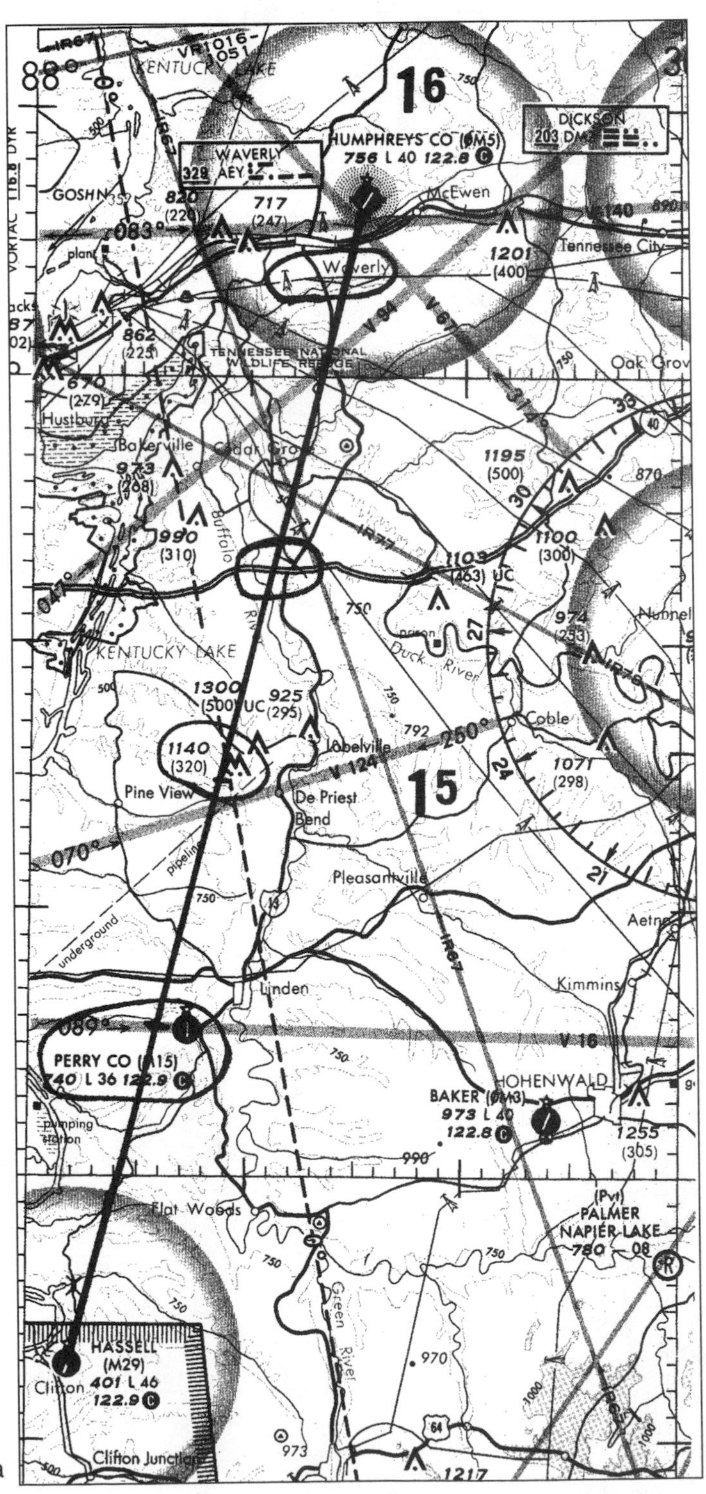
16
HUMPHREYS CO (0M5)
756 L 40 122.8
DICKSON
WAVERLY
328 AEY
GOSHN
820
(220)
717
(247)
083°
McEwen
Tennessee City
1201
(400)
Waverly
862
(225)
TENNESSEE NATIONAL WILDLIFE REFUGE
Oak Grove
670
(279)
Hustburg
Bakerville
Cedar Grove
973
(268)
1195
(500)
870
990
(310)
Buffalo
1100
(300)
1103
(463) UC
KENTUCKY LAKE
Duck River
974
(253)
1300
(500) UC
925
(295)
1140
(320)
Lobelville
792
250°
Coble
V 124
15
1071
(298)
Pine View
De Priest Bend
070°
pipeline
Pleasantville
underground
Aetna
Linden
Kimmins
089°
V 16
PERRY CO (1M15)
740 L 36 122.9
HOHENWALD
BAKER (0M3)
973 L 40
122.8
1255
(305)
990
Flat Woods
(Pvt)
PALMER
NAPIER LAKE
780
Green River
HASSELL
(M29)
401 L 46
122.9
Clifton
970
Clifton Junction
973
1217
Fig. 8.2a

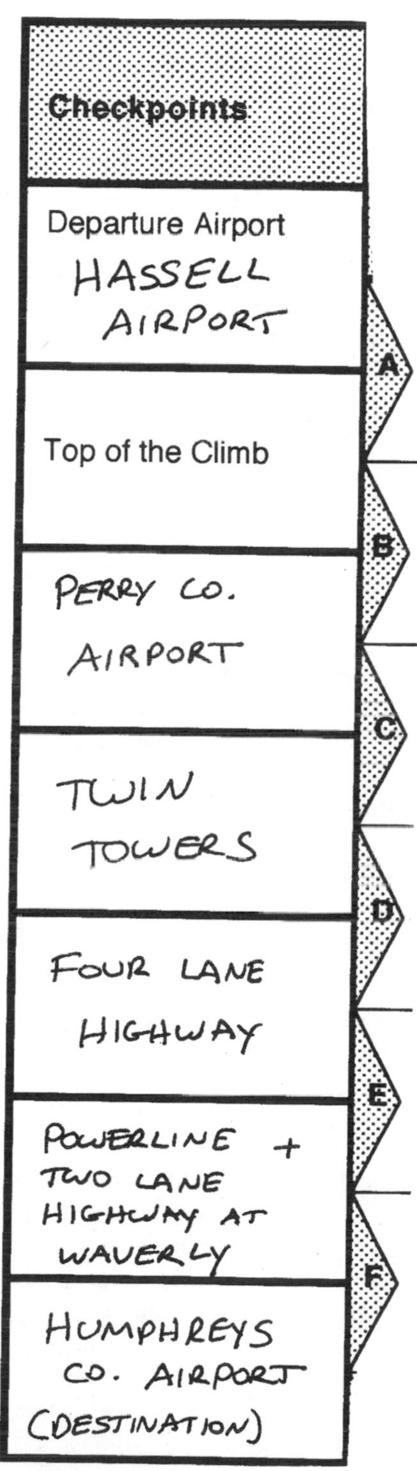

Checkpoints
Departure Airport
HASSELL AIRPORT
A
Top of the Climb
B
PERRY CO. AIRPORT
C
TWIN TOWERS
D
FOUR LANE HIGHWAY
E
POWERLINE + TWO LANE HIGHWAY AT WAVERLY
F
HUMPHREYS CO. AIRPORT (DESTINATION)

Fig. 8.2b

points that lie as close under the True Course Line as possible.

Also from Chapter 4 the mileage scales on the Navigation Plotter were discussed. Use the Plotter and the correct scale to measure the mileage for the entire flight and between each Checkpoint. When measuring from one Checkpoint to another it is best to make a small "hash" mark across the True Course Line to indicate where the measurement has stopped. That same hash mark can be used to begin measuring to the next Checkpoint.

Column 17 of the Navigation Record (figure 8.1) is the place to record the mileage. At the very top of the column is a non-shaded box titled "Total Miles." Write in the total nautical miles for the flight. In all boxes down the page under column 17 there is a diagonal line. On the upper side of the diagonal line is where the "Leg" mileage is placed. The Leg mileage is simply the distance from one Checkpoint to another. In the lower side of the diagonal line is the "to go" information. This is the total miles left to go on this flight before reaching the destination airport. Total miles minus the Leg equals the "to go" distance.

The Navigation Record is offset. Look at the Checkpoint boxes (figure 8.1). Each box has an arrow pointing to the right and into the rest of the Navigation Record. Each arrow is lettered "A" through "G." All the information contained in the boxes to the right of the arrow will pertain to that leg. For instance, all the information in the boxes along the row marked by the arrow "B" would be for the leg that begins after the cruising altitude is reached (Top of Climb) and the first landmark selected as a Checkpoint.

The Navigation Record's lettered rows and its numbered columns form a grid. For example, the position 16A would therefore be on the top row under the label "CH." Position 17A is a distance or mileage box, but this particular mileage box pertains to the mileage covered in the climb because row "A" is labeled "Enroute Climb." The enroute climb calculation is covered in Chapter 9.

All other mileage between Checkpoints does not come from a calculation, but rather from direct measurement on the chart. The mileage is measured between the hash marks of each checkpoint, and the miles are recorded on the upper side of each box under column 17. The "Total Miles" minus the distance of a Leg equals the "to go" distance. Starting from the top with the "Total Miles" box the legs can be subtracted and the "to go" is carried to the next lower box, until reaching the destination where "to go" will be zero.

In rows B through G the mileage between each leg is recorded in the upper side of each box's diagonal line. Now we can use the Ground Speed that would have already been calculated (Column 9) and determine how long, in minutes, it will take to travel between each Checkpoint and to the destination. The minutes between each Checkpoint will be recorded for each Leg in column 18: "Estimated Time to Checkpoint."

HOW MUCH TIME?

To determine the length of time to travel a known distance with a certain Ground Speed we use the Calculator side of the Flight Computer. On most Flight Computers there is a "code" to help the pilot determine just how to use the Calculator Side. Figure 8.3 displays a sample code. The code can usually be found on the face of the Calculator Side.

You can see that there is a *Pyramid* shape on the code. This Pyramid acts as a pointer. The pilot turns the wheel of the Calculator Side, so that the Pyramid points directly at the Ground Speed. While the Pyramid is aimed at the GS the "Time" can be read off the computer directly across from the mileage. "Time" has two scales. The "minutes" scale is on the rim of the wheel that turns. The "hours" scale is also located on the movable wheel, but farther in from the edge of the wheel. The Ground Speed that the pyramid points out is not on the movable wheel, but instead on the farthest outside scale. Distance is also located on this most outside scale.

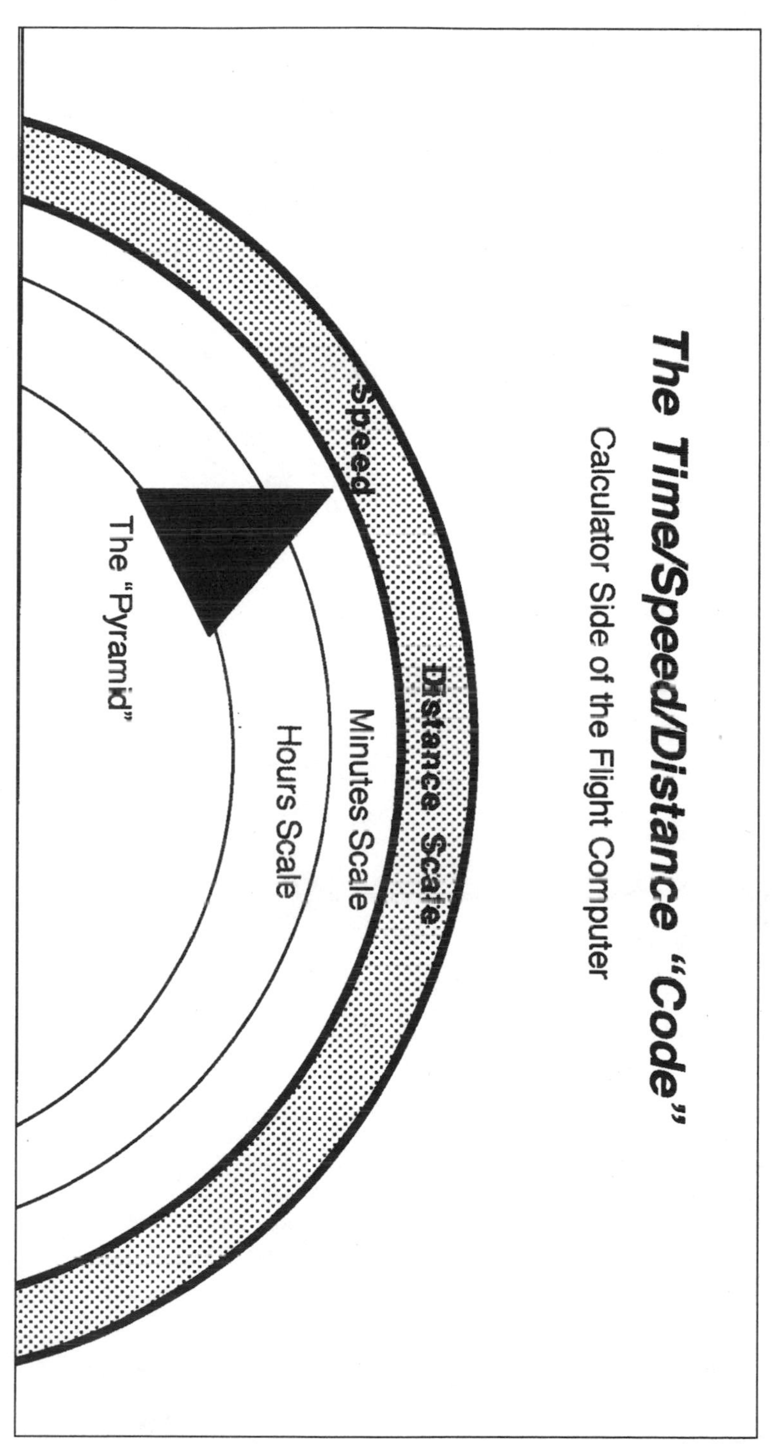
The Time/Speed/Distance "Code"
Calculator Side of the Flight Computer
Speed
Distance Scale
Minutes Scale
Hours Scale
The "Pyramid"

Fig. 8.3

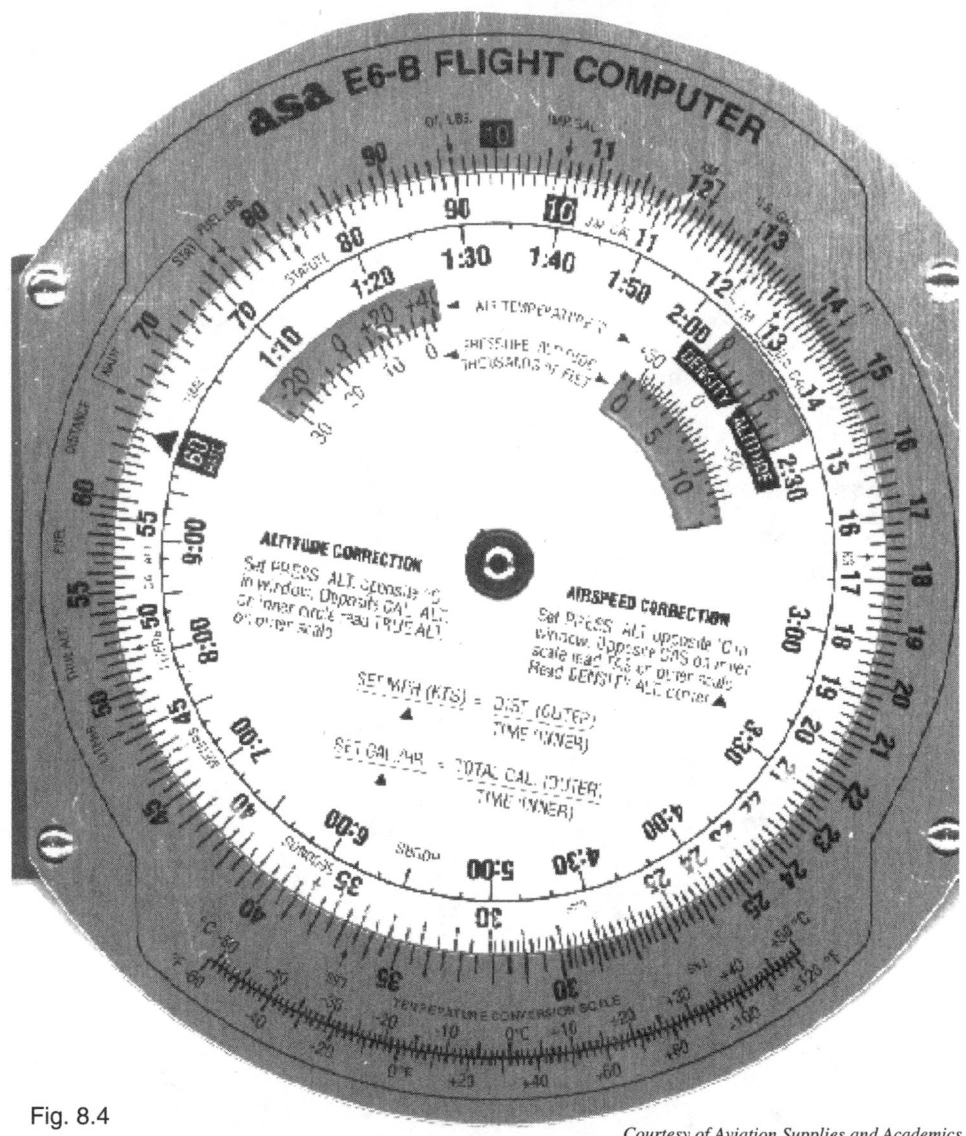

Fig. 8.4

Courtesy of Aviation Supplies and Academics

Figure 8.4 illustrates the Calculator Side. Notice the position of the Pyramid and the Ground Speed it is pointing out. With the wheel in this position, and the airplane at this Ground Speed, how long will it take the airplane to travel a distance of 50 miles? 150 miles? 5 miles? 500 miles?

The answers to the Speed/Distance/Time problems from the Navigation Record columns 9 (Speed) and 17 (Distance) are recorded in column 18 (Time).

HOW MUCH FUEL?

Now the most important calculation: Fuel needed for the flight. Nothing could be more serious to the pilot than to run out of fuel in the air, so this next step is crucial. The amount of fuel the airplane will burn on a flight depends on two factors. First, the time the airplane's engine is operating, and second, the speed in which the fuel is burned when the engine is operating. The "Time" factor was calculated in Navigation Record column 18. The second factor is the fuel rate.

Automobiles usually consider "gas mileage" and calculate the "miles per gallon." Airplanes usually do not figure "miles per gallon," but instead calculate the **Gallons per Hour (GPH)**. Since the Legs are all timed, it is easy to think of fuel in terms of time as well. The "fuel flow" or GPH is a number that first comes from the airplane's Pilot Operating Handbook. The book will have a table of different GPH rates based on various conditions. Figure 8.5 is a sample Fuel Consumption table.

The GPH rate depends on three factors: Pressure Altitude, Temperature, and the speed the pilot operates the engine: RPM. The Pressure Altitude used here is the same as discussed before. Pressure Altitude is the Altimeter setting when it is set to 29.92. But there is a problem. The Pressure Altitude the pilot might read on the ground during pre-flight planning will not be the same as the Pressure Altitude when the airplane is in the air and at the final cruising altitude. The pilot must note the difference between the MSL altitude (airport elevation) of the airplane while on the ground and the Pressure Altitude, then apply this difference for use with higher altitudes. For instance, if the airport elevation is 500 feet and the Pressure Altitude is read to be 1,000 feet, then there is a 500 foot difference. If the pilot plans to fly a cruising altitude of 5,500 feet then a cruising Pressure Altitude of 6,000 feet would be used on the Fuel Consumption table. Often this table will need to be interpolated.

The next factor is temperature. The air temperature usually gets colder with an increase in altitude. The "standard" sea level temperature is 59°F which is the same as 15°C. The standard "cool off rate" of the air with an increase in altitude is 3.5°F, which is the same as 2°C. This cool off rate is called the *Standard Lapse Rate.*

The following table outlines the lapse rate from sea level

Altitude	Standard Temperatures
Sea Level	15° C
1,000	13° C
2,000	11° C
3,000	9° C
4,000	7° C
5,000	5° C
6,000	3° C
7,000	1° C
8,000	-1° C
9,000	-3° C
10,000	-5° C
11,000	-7° C
12,000	-9° C

Across the top of the Fuel Consumption table (figure 8.5) are listed the headings "20°C Below Standard," "Standard Temperature," and "20°C Above Standard." The column used to determine GPH depends on what the "at altitude" temperature is. This "at altitude" temperature can be estimated by using the Standard Lapse rate beginning with the surface temperature and subtracting as you go up, or the weather briefer at a Flight Service Station can give "Temperatures Aloft" (more on FSS weather briefings in Chapter 14).

If the temperature was 23°C at 6,000 feet, which GPH column should be used? The pilot should use the "20°C Above Standard" column. The Standard temperature at 6,000 is 3°C and 23°C is obviously 20° greater.

Last, the RPM setting. The pilot should consult the manufacturer's recommendations to determine what is the most efficient setting with the least wear on the engine.

Referencing figure 8.5, the following GPH rates calculate as:

Press Alt	Temperature	RPM	GPH
1. 4,000	7°C	2,400	4.1
2. 8,000	-21°C	2,300	3.6
3. 3,000	9°C	2,500	*4.75
4. 6,000	13°C	2,400	*3.8
5. 7,000	-9°C	2,300	*3.55

**interpolation required*

When the GPH rate has been determined for a particular flight, that number should be placed in the top of the Navigation Record column 19, labeled "GPH." To determine just how much fuel will be burned while the airplane is cruising at a level altitude use the "Time" from column 18 and the "GPH" at the top of column 19 and the Calculator Side of the Flight Computer.

Fuel Consumption Table

Press Altitude	RPM	20 C Below Standard	Standard Temperature	20 C Above Standard
2,000	2600	6.0	5.5	5.1
	2500	5.3	4.9	4.6
	2400	4.7	4.3	4.1
	2300	4.1	3.9	3.7
	2200	3.7	3.5	3.3
4,000	2600	5.6	5.2	4.8
	2500	5.0	4.6	4.3
	2400	4.4	4.1	3.9
	2300	3.9	3.7	3.5
	2200	3.5	3.3	3.2
6,000	2600	5.2	4.8	4.5
	2500	4.7	4.3	4.1
	2400	4.2	3.9	3.7
	2300	3.7	3.5	3.4
	2200	3.5	3.3	3.2
8,000	2600	4.9	4.6	4.3
	2500	4.4	4.1	3.9
	2400	4.0	3.7	3.5
	2300	3.6	3.4	3.2

Fig. 8.5

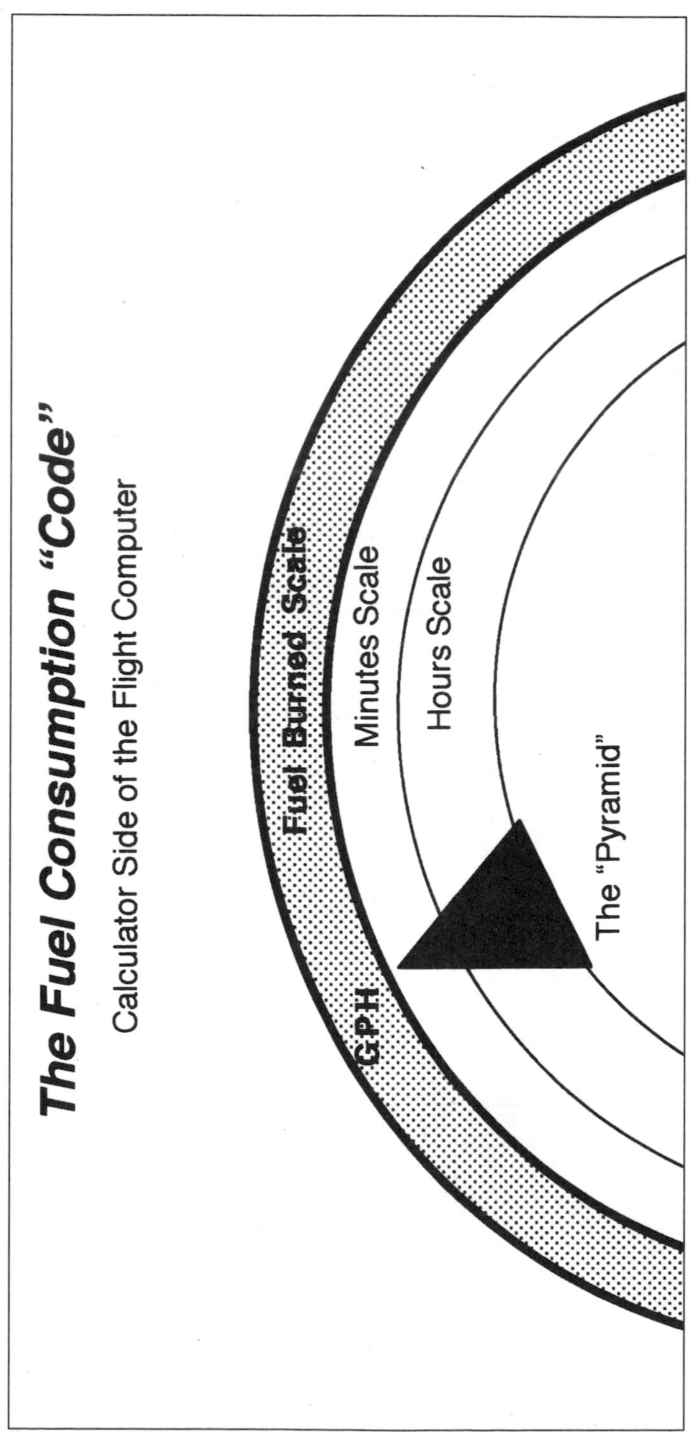

Fig. 8.6

The Calculator Side will have another "code" to use in determining "Fuel Burned." Figure 8.6 is a sample code for Fuel Consumption. In this instance the Pyramid is pointed at the GPH rate that was discovered on the Fuel Consumption table. With the Pyramid in this position the Fuel Burn can be read across from the minutes scale.

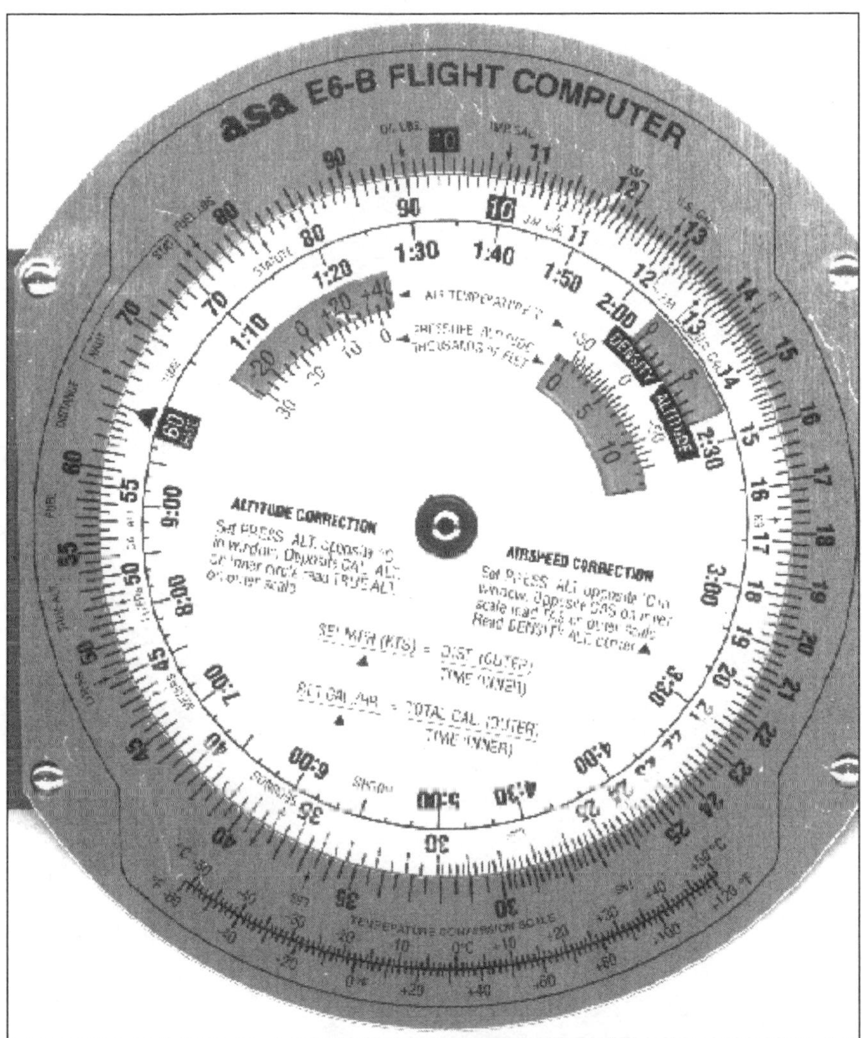

Fig. 8.7 *Courtesy of Aviation Supplies and Academics.*

Figure 8.7 is an illustration of the Calculator Side with the Pyramid pointed at a GPH rate. At this rate how much fuel will this airplane burn in 30 minutes? in 1 hour? in 17 minutes? in 2 hours and 50 minutes?

These Fuel Burn amounts for each leg should be recorded in column 19. The position of these numbers should be on the upper side of the diagonal lines. The lower side of the diagonal lines will be used later to calculate the amount of fuel remaining in the airplane's fuel tanks.

Try some practice problems with your own flight computer and verify these numbers:

Sample Time/Speed/Distance Problems

	SPEED (kts)	DISTANCE (nm)	TIME
1.	95	30	______
2.	110	22	______
3.	90	9	______
4.	105	140	______
5.	88	1.1	______
6.	______	12	9 minutes
7.	______	25	15 minutes
8.	______	200	1 hr 45 minutes
9.	______	0.8	30 seconds
10.	82	____	30 minutes

Sample Fuel Consumption Problems

	GPH	TIME	Fuel Burned (gal)
1.	6.4	15 minutes	______
2.	10.0	12 minutes	______
3.	5.1	10 minutes	______
4.	6.6	5 minutes	______
5.	7.2	2 hrs 30 mins	______
6.	8.5	3 hrs 10 mins	______
7.	______	1 hr 15 mins	9 gallons
8.	______	30 minutes	9 gallons
9.	______	1 hr 30 minutes	13 gallons
10.	______	10 minutes	1.5 gallons

Answers:

Time/Speed Distance Problems

1. 19 minutes
2. 12 minutes
3. 6 minutes
4. 1 hour 20 minutes
5. .75 minutes or 45 seconds
6. 80 kts
7. 100 kts
8. 114 kts
9. 96 kts
10. 41 nm

Fuel Consumption Problems

1. 1.6 gallons
2. 2.0 gallons
3. .85 gallons
4. .55 gallons
5. 18 gallons
6. 27 gallons
7. 7.2 GPH
8. 10.8 GPH
9. 8.7 GPH
10. 9.0 GPH

9

Aircraft climb performance and fuel reserves

IN CHAPTER 8 THE TOPIC OF DISTANCE, TIME, AND FUEL BURN WERE discussed. You may have noticed however, that these numbers were calculated at a "cruise" power setting. The problem is that airplanes do not just start up and start cruising! Airplanes must climb to a predetermined altitude, level off, and only then take advantage of fuel savings. Figure 8.5, the Fuel Consumption table, is a list of "cruise" fuel rates that assumes the pilot has reached altitude and has properly leaned the engine. But, to consider the real situation we must examine what is taking place in the climb as well.

Which condition would burn the greater amount of fuel?

1. A driver at a stop light quickly pushes the gas pedal all the way to the floor when the light turns green and accelerates to 65 miles per hour in 10 seconds, or
2. A driver while traveling at 65 miles per hour sets the "cruise control" and steps off the gas pedal.

Obviously the first situation would burn more fuel than the second. Think of the airplane's climb like the first situation and the airplane's cruise like the second. The climb in a small airplane will have a slower Ground Speed, and burn more fuel than the remainder of the flight. But if the pilot calculates the Ground Speed, time to destination, and fuel burn as if the airplane were cruising the whole time, you can see that an error would be made.

How much error is made if the climb is ignored? With small, single-engine airplanes that do not climb very high in the first place, the errors are not great, but most readers of this book do not want to fly small, single-engine airplanes their whole life. Most want to graduate to bigger, faster, and more fuel-drinking planes. The idea is to understand how to calculate the climb performance now, when the numbers are smaller, so that it will be easy when the airplane and numbers are larger!

CLIMB INFORMATION

Looking back at figure 8.1, The Navigation Log, you can read that the horizontal row with the lettered arrow "A" has the words "Enroute Climb" in the Altitude box. *This means that all information in these boxes across the Navigation Log on the horizontal "A" row will pertain to the airplane's climb.* What items will be different during climb than in the eventual cruise? First, the Indicated Airspeed will be slower. That will make CAS, TAS and Ground Speed slower as well. When GS is slower there may be additional Wind Correction Angle needed, because a slower airplane will be pushed around by the wind easier. If the climb changes GS and WCA, then True Heading and ultimately Compass Heading will be different in the climb than from cruise. The Distance to the *Top of the Climb* (the first Checkpoint) is determined by climb performance. The time to the Top of the Climb and the fuel burned while climbing both must come from information about the airplane's climb performance.

Figure 9.1 is a sample "Climb Performance Chart." A chart like this will be included in your airplane's Pilot Operating Handbook (POH). On the far left of the chart is listed the airplane's weight. This chart simply lists the airplane's maximum takeoff weight. The next column over is the pressure altitude levels.

On the far right the chart is divided into three columns: 1. Distance to climb in nautical miles, 2. Time to climb in minutes, and 3. Fuel burned during the climb in gallons. These three columns also are marked "Nav Record 17A," "Nav Record 18A," and Nav Record 19A." These Nav Record numbers correspond to the proper boxes on the Navigation Log where the numbers from this climb chart will be recorded.

The chart helps pilots get the most from their airplanes. If a pilot wants to get the best performance from the airplane during a climb from sea level to 4,000 feet pressure altitude then he or she should hold the pitch of the airplane during

Climb Performance Chart

Weight	Press Alt Feet	Climb IAS Knots	From Sea Level Distance N.Miles Column 17	From Sea Level Time Minutes Column 18	From Sea Level Fuel Burn Gallons Column 19
1,800 Pounds	Sea Level	68	0	0	0.0
	1,000	68	2	2	0.2
	2,000	67	4	3	0.5
	3,000	66	6	5	0.7
	4,000	65	8	7	1.0
	5,000	65	10	9	1.3
	6,000	64	13	11	1.6
	7,000	64	16	14	1.9
	8,000	63	19	17	2.3
	9,000	63	23	20	2.7
	10,000	62	27	23	3.2

1. Pilots should ADD 0.8 gallons to allow for Start, Taxi, and Run-up.
2. Distances shown here in Nautical Miles are calculated with **Zero Wind.**

Fig. 9.1

the climb so that the Indicated Airspeed (IAS) will read 65 knots. At this angle, 65 knots, the airplane should travel 8 nautical miles over the ground while climbing from sea level to 4,000. The chart shows that it will take the airplane 7 minutes to reach 4,000 from Sea Level, and the engine will burn 1.0 gallon during this full power climb. These numbers (8 nautical miles, 7 minutes, and 1.0 gallon) should be recorded on the Navigation Record at positions 17A, 18A, and 19A.

CLIMB PROFILE

Sometimes climbs must be made from airports that do not happen to have a Pressure Altitude of sea level. The chart can also be used in those cases. A pilot is planning a climb from a takeoff Pressure Altitude of 2,000 to a cruising Pressure Altitude of 8,000, for example. How much distance, time, and fuel will it take? To answer the question you must look *inside* the chart. If the pilot had started at sea level the climb to 8,000 would cross 19 miles. But in this example the pilot is not at sea level, but is already up at 2,000. A climb from Sea Level to 2,000 would take 4 miles, so the difference between 2,000 and 8,000 will be the difference between 4 miles (sea level to 2,000) and 19 miles (Sea Level to 8,000) or 15 miles. Using the same calculation you can see that from 2,000 it will take the airplane 14 minutes to reach 8,000 and the engine will burn 1.8 gallons. Refer to figure 9.2 for an illustration of this climb and the numbers produced.

VFR Cruising altitudes are usually at *500 foot levels*. This means that interpolation of the chart will be needed. What would the numbers be for this airplane to climb from 1,000 feet to 4,500 feet? To answer you must "split the difference" between the 4,000-foot and 5,000-foot numbers, and then subtract off the 1,000-foot numbers. Distance would be 7 miles, in 6 minutes, while burning 1.0 gallons (rounded up).

You might notice that no mention has been made of a descent and the change in performance while coming back down. In this book no attempt will be made to predict the descent performance. Theoretically, the GS should be faster, and the fuel burn rate slower, but descents in actual flying are hard to predict and the savings small. I prefer to overestimate my need for fuel. If I do save fuel in the descent, I will just have extra reserve fuel.

Using the Climb Performance Chart (figure 9.1) and the Fuel Consumption table (figure 8.5) the entire Navigation Record can be calculated, including the climb and the cruise. As stated many times before the fuel amount is critical to the safety of the flight and is one of the biggest reasons why pilots should take the time to fill out a Navigation Record in the first place. After calculating all the speeds, and fuel burn numbers, there still remains a fundamental question:

How much fuel do I need for the flight? The answer to this question is broken down into four parts:

1. The fuel for engine start up, taxi, and run-up.
2. The fuel for climb to altitude.
3. The fuel for cruise to the destination.
4. The legal fuel reserves.

Number 1. The allowance for Start, Taxi, and Engine run-up is included in the notes at the bottom of the Climb Performance Chart (Figure 9.1). This is an average amount of fuel. Clearly a pilot will not burn 0.8 gallons before each takeoff. Sometimes the pilot is required to taxi a long distance at a large airport, others the taxi distance is short from the parking area. There will be times at busy airports when the pilot must wait behind other planes for several minutes before takeoff, and other flights will have no delay at all. The 0.8 gallons should be used however, unless you believe a greater allowance should be figured for expected delays.

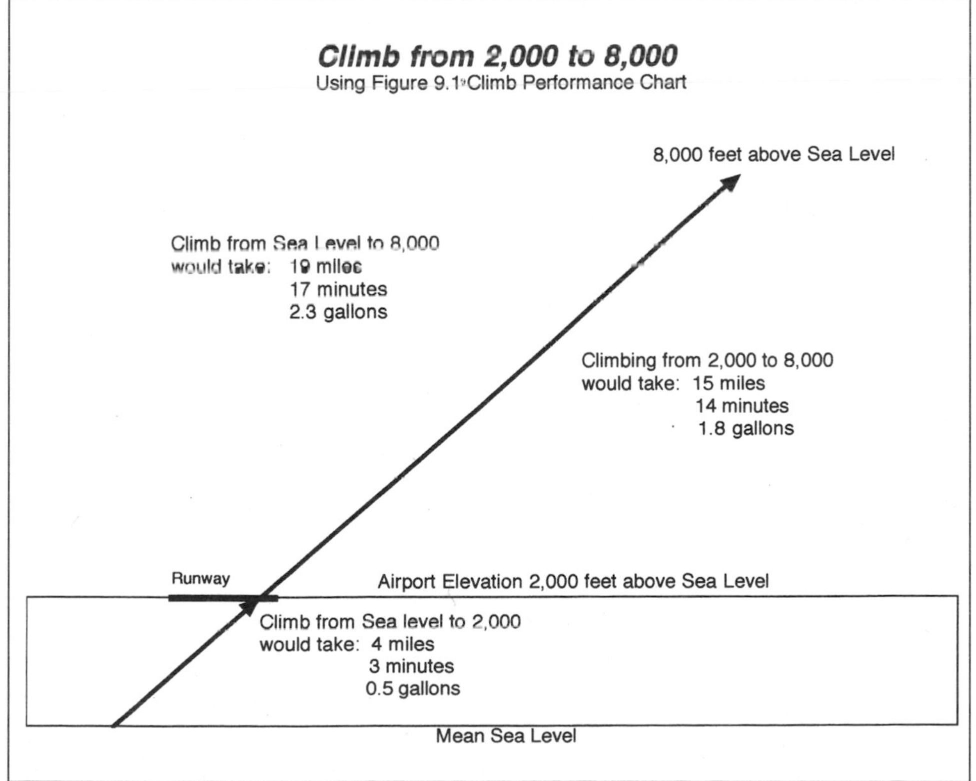

Fig. 9.2

Number 2. The fuel to climb to altitude comes directly from the Climb Performance Chart.

Number 3. The fuel to cruise to the destination comes from adding together the fuel burn from all level flight legs of the trip.

Number 4. The FAA regulations require that VFR pilots be able to fly to the airport of intended landing plus a fuel reserve of 30 minutes for daylight flights and 45 minutes in the dark. This reserve is calculated at the Cruise fuel burn rate (GPH) that is found on the Fuel Consumption table.

When all is added together a total fuel required amount is the result. Is the fuel required for the flight (numbers 1, 2, 3 and 4) equal to or less than the amount of fuel in the airplane? If the airplane has more fuel than required you are ready to go. If the airplane does not have the calculated requirement of fuel then this trip requires a stop for more fuel!

I would always recommend that for small, single-engine airplanes, cross country flights be started with ***full tanks!*** Yes, someday you may fly airplanes that will depart with less than full tanks for reasons of loading or just economy, but with light "trainer-type" airplanes — FILL UP. Do not guess how much fuel is in the tanks by using the fuel gauges or seeing if your finger can reach the fuel down in the tank. The only way you can know for sure how much fuel is in the airplane is by topping off.

Verify these sample problems with climb performance:

Climb from	to Cruise Altitude	How many Minutes?	Miles?	Gallons?
1. Sea Level	9,000	20	23	2.7
2. 2,000	7,000	11	12	1.4
3. 1,000	8,500	16.5	19	2.3
4. 500	5,000	8	9	1.2
5. 1,500	6,500	10	11.5	1.4

10
Which altitude should be flown?

THE EXACT CRUISING ALTITUDE THAT SHOULD BE FLOWN IS NOT JUST a guess. For every flight there should be a strategy to determine the best altitude. The strategy should be based on four elements: 1) the elevation of the terrain that will be flown over, 2) the wind, 3) the fuel consumption of the airplane, and 4) the cruising altitude regulations.

The first and most obvious factor is the terrain that must be crossed on the way to the destination. The pilot must select a cruising altitude that will be higher than anything along the way. Sometimes the terrain is higher than the airplane can climb. In that case, the pilot must fly *around* not *over* the terrain. The pilot should consider obstructions as well as ground elevations in the decision. Look over the True Course Line that is drawn on the chart for your particular trip. On the Sectional chart the Latitude and Longitude lines form boxes all over the chart. In each box there will be a blue "number set" that represents the highest elevation and obstructions within that box. The pilot must plan the flight at an altitude that is higher than the number sets along the flight path.

After the pilot determines the lowest safe cruising altitude the question changes to: What is the *best* altitude? The best altitude would be the altitude that would take the airplane to the destination in the least time and burning the least fuel. The time it takes to reach the destination will be determined by the Ground Speed and this is determined by the wind. Very often the wind speed will increase with altitude. This means that if a proposed flight is lucky enough to have a tailwind, the pilot would ordinarily want to go very high so as to take advantage of the greatest tailwind. But if the flight had a predicted headwind, the pilot would want to stay low and away from the slower ground speed that would exist at a higher altitude.

Pilots receive their cruising altitude wind information from what is called the ***Winds and Temperatures Aloft Forecast***. The computer code for this forecast is FD. When a pilot calls the Flight Service Station to receive weather information, the briefer will read from the FD report. For pilots who use the ***Direct User Access Terminals*** or DUATS to receive information the FD can be read directly off of the monitor.

The following is a sample FD report:

	3,000	6,000	9,000	12,000	15,000	18,000
BNA	1505	1615+12	1830+09	1842+05	1955+01	2065-03
DIA			2537-05	2655-11	2880-15	8005-17
RDU	9999	0305+16	0512+14	0518+12	0622+10	0625+08
STL	0912	0920+08	1128+02	1240-02	1260-05	1475-10

The FD report is easy to decode. The first two digits are the wind direction rounded off to the nearest 10 degrees. This means that "03" is 030°, "18" is 180°, and "22" would be 220°. The second two digits are the wind's velocity in knots. The last two digits are the Celsius temperature. The code, "1215+10" is read: Wind from 120°, at 15 knots, and the temperature at that altitude is +10° Celsius.

All altitudes are given based on Mean Sea Level altitudes (MSL) and **wind is not reported within 1,500 feet of the ground.** That is why there are no winds reported from Denver International Airport (DIA) at 3,000 and 6,000 feet. The elevation of Denver (the "mile high" city) is above 5,000 feet. There can be no winds at 3,000 feet at Denver, because 3,000 MSL is about 2,000 feet *under ground!*

INTERPRETING THE WIND REPORT

The code from DIA at 18,000 feet reads "8005-17." The first two digits are the wind directions, but there is no such direction as 800°, so is this entry a misprint? No, whenever the wind direction is given with an impossible direction it is a signal to the pilot about a *code within a code!* The second two digits are the wind velocity, but what if the wind speed becomes faster than 100 knots? How can a number with three digits fit into a code that only allows two? To solve this problem the "code within a code" was developed. Whenever a direction number appears that is actually not a direction (anything greater than 360°), this means that wind speed has passed 99 knots. To decode, take the first two digits and subtract 50 to get the actual wind direction and add 100 to the second two digits to get the actual wind speed. A listing of "8005-17" decodes to read "Wind 300° (80-50=30) at 105 knots (05+100=105) with a temperature of -17°C.

COMMON SENSE ALTITUDE SELECTION

When making altitude plans the pilot should look at the wind direction and velocity trend from the ground up. If a pilot were planning a cross-country flight near Nashville (BNA) and the direction of travel was to the South, what altitudes would be best? From the sample FD report it looks as if the pilot would need to stay low. The higher the pilot would climb the faster the head-winds would become and the slower the Ground Speed would become. What about a flight to the West from St. Louis (STL)? This pilot would want to go high in order to take advantage of the tailwinds coming from the East.

After considering the terrain, obstructions, and wind, the pilot now determines altitude based on the airplane's fuel consumption. Look back at figure 8.5, the Fuel Consumption table. Looking across this table, it is easy to see that the higher you fly the less fuel you burn. The actual numbers on fuel consumption tables like figure 8.5 are based on the proper "leaning" of the airplane's engine. As the airplane climbs away from sea level the air gets thinner. Since the engine runs on a precise ratio of air and fuel, when the air density is less, the fuel flow should also be less in order to maintain the proper ratio. This means that air-plane engines can "sip" not "guzzle" fuel at high altitudes. Talk to your instructor and read the airplane's POH to determine the exact procedure to lean the engine.

All things otherwise being equal, a pilot would prefer to be up high and take advantage of lower fuel rates (GPH). But climbing to a higher altitude comes at a price. Look back once again at figure 9.1, the Climb Performance Chart. Climbs to altitude burn more fuel.

On figure 9.1 you can see that a climb from sea level to 8,000 feet would take 17 minutes and burn 2.3 gallons of fuel. If this climb fuel rate were continued for one hour the rate would be 8.1 GPH. Figure 8.5 shows an 8,000-foot cruise with 2,500 RPM and standard temperature is 4.1 GPH. In other words, the air-plane burns about twice as much fuel while climbing as it does while cruising. The pilot must *pay* for the climb, but does the resulting high altitude cruise savings pay the pilot back and then some? The answer depends on how long the airplane will be able to stay high while enjoying the fuel savings.

If a pilot burns 3 gallons of fuel just to get high enough to save 2 gallons, then the climb was not a good investment. If the pilot burns 3 gallons in the climb, but then, because of the length of the trip, stays high for enough time to save 8 gallons, then the climb was a good idea.

When the pilot determines altitude after considering terrain, obstructions, wind, and airplane performance, there is yet one more factor: The FAA's cruising altitude regulations. The Federal Aviation Regulations require that airplanes flying cross-country VFR use the following altitudes:

For ***Magnetic Courses*** of 360° to 179° use Odd-numbered altitudes plus 500:

3,500 5,500 7,500 9,500 11,500 13,500 15,500 17,500

For ***Magnetic Courses*** of 180° to 359° use Even-numbered altitudes plus 500:

4,500 6,500 8,500 10,500 12,500 14,500 16,500

No VFR flights are allowed above 18,000 feet, so no VFR altitudes are at those altitudes, and the cruising altitudes do not take effect until above the ground by higher than 3,000 feet. Also make special note of the fact that the difference between odd and even thousand-foot levels plus 500 is based on ***Magnetic Course***. The magnetic course is the True Course corrected for Variation. In Chapter 5 of this book the examples used started with True Course, then allowed for wind correction to get True Heading. However, for altitude determination you must look at True Course then Variation first to get Magnetic Course. The Magnetic Course is the actual path across the ground. An airplane with a crosswind could have an MC of 175 and at the same time a TH of 185. This could cause some confusion when selecting an altitude.

In 1985 this confusion apparently lead to a fatal mid-air collision between a Piper and a Mooney. The Piper had an MC of 173° and therefore should have flown the trip at an odd thousand plus 500-foot level, but the wind was from the West and the Piper had a "crab" angle which placed the *Heading* at about 185°. The Mooney had an MC of 347° and was correctly flying an even thousand plus 500-foot level. The collision occurred at 4,500 feet. Did the pilot of the Piper incorrectly use Magnetic Heading instead of Magnetic Course to determine his cruising altitude?

1. The altitudes used for VFR flying above 3,000 AGL are at thousand plus 500 foot levels (4,500 7,500 13,500 etc.) but the winds given in the Winds and Temperatures Aloft Forecast are only at thousand-foot levels. This means that there will be times when pilots must interpolate the FD report. What would be the winds for a VFR flight at 4,500 feet in this example?

3,000	6,000
1512	1718+15

Since 4,500 feet is halfway between 3,000 and 6,000, the pilot must calculate "halfway" for the wind's direction and velocity. In this example the wind at 4,500 feet would be 160° at 15 knots. Usually for altitudes within 500 feet of the FD reported altitude level, it is acceptable to use the FD numbers. For instance, when computing the winds for 5,500 feet, the information for 6,000 feet can be used.

2. You are planning a VFR cross-country flight to an airport that is 175 nautical miles away. The terrain to be crossed is never higher than 2,000 feet MSL, but your airplane cannot climb above 10,000 feet. By using charts, a plotter, and a flight computer you have determined the following information:

CAS = 87 knots　　Variation – 5°W
TC = 241°　　Deviation = 0°

By using a DUAT computer you receive the following FD report for the area of your flight:

3,000	6,000	9,000
1505+15	0918+08	0722-01

Reference the following table of information, and then after considering the terrain elevation, the cruising altitude regulations, airplane performance (figures 8.5 and 9.1) and the wind, determine the ***BEST*** altitude for the flight.

	Wind Direction	Wind Velocity	Temp	TAS	WCA	GS
3,000						
4,500						
6,000						
7,500						
9,000						

The *best* altitude would be 8,500 feet! Must be an even thousand with a Magnetic Course of 246°. The fastest GS and better GPH would be 8,500. The climb to 8,500 from sea level will cost 2.5 gallons, but only take 23 miles of the trip, leaving over 150 miles to enjoy a better GPH at 8,500 feet.

Answers:

	Wind Dir.	Wind Vel.	Temp	TAS	WCA	GS
3,000	150	05	+15	92	3L	92
4,500	120	12	+12	94	7L	99
6,000	090	18	+08	96	6L	111
7,500	080	20	+04	98	4L	116
9,000	070	22	-01	100	3L	121

11 The complete Navigation Record

IT HAS TAKEN 10 CHAPTERS TO THIS POINT, BUT ONLY NOW CAN WE proceed to do the complete Navigation Record. Figure 8.1 is the Navigation Record. In fact, the Navigation Record is not totally completed until the flight has been completed, but in this chapter we will complete everything under the Navigation Record's heading "Pre-flight Calculations and Planning." For this flight use the Sectional Chart provided in figure 11.1. The Flight will begin at the Upper Cumberland Regional Airport (SRB) located at 36° 03'N. Lat and 85° 32' W. Long. The flight will end at the Bowling Green-Warren County Regional Airport (BWG) located at 36° 58' N. Lat. and 86° 25' W. Long. The flight will be made in VFR weather conditions. The pilot calls the nearest FSS by dialing 1-800-WX-BRIEF (this works nationwide). The briefer reports that "good VFR exists over the route of flight to Bowling Green and is expected to stay that way until long after arrival at the destination." The clouds are 8,000 scattered, visibility is better than 10 miles. The winds and temperatures aloft for the area of the flight are given as follows:

3,000	6,000	9,000
1814+19	2420+13	2826+07

TRUE COURSE AND CHECKPOINTS

Draw a True Course Line between the two airports. Ordinarily the pilot can select any checkpoints they like, but for this example use the checkpoints that have been pre-selected. For calculations assume that the pressure altitude is equal to the MSL altitude.

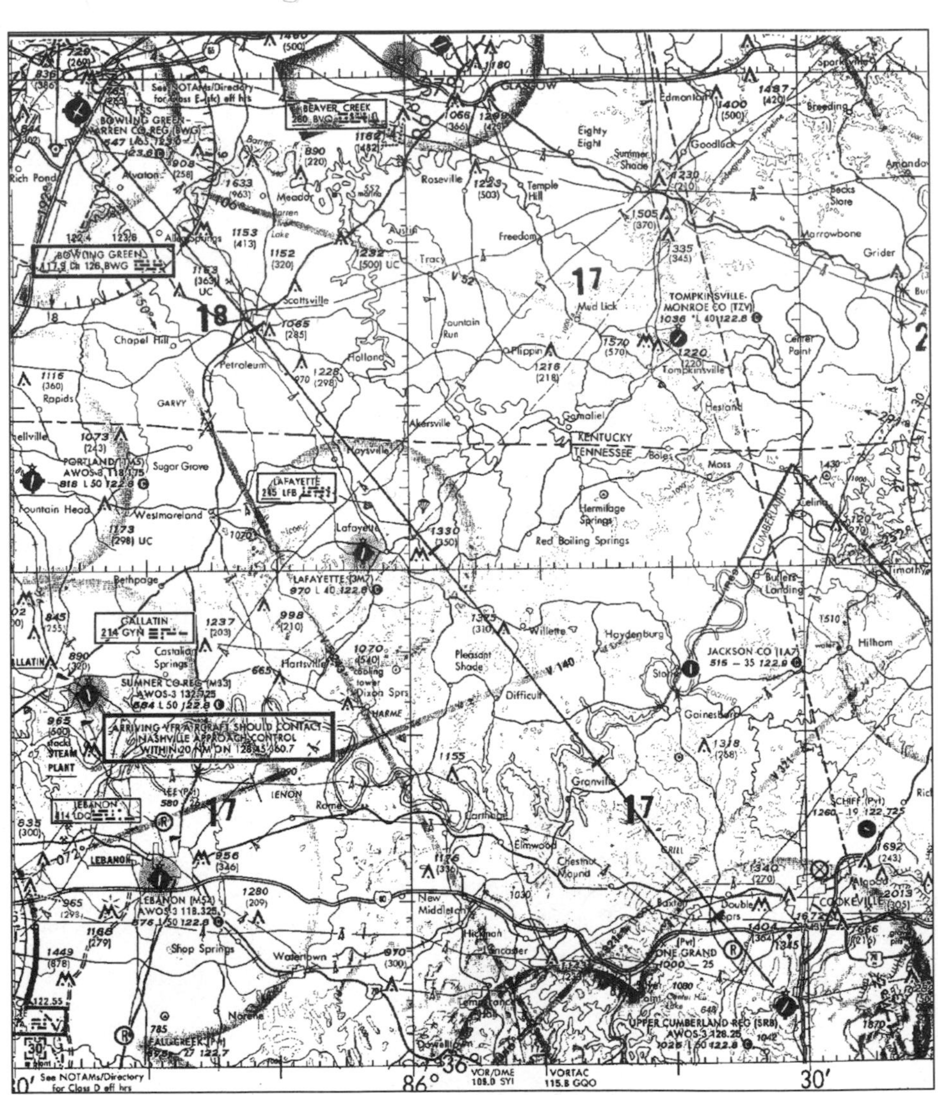
BEAVER CREEK
Eighty Eight
Edmonton
Goodluck
Summer Shade
Roseville
Temple Hill
Freedom
Becks Store
Marrowbone
Grider
Rich Pond
Alvaton
Meador
Scottsville
Tracy
Mud Lick
TOMPKINSVILLE-MONROE CO (TZV)
Tompkinsville
Center Point
Chapel Hill
Petroleum
Holland
Fountain Run
Flippin
Rapids
Akersville
Gamaliel
Hestand
KENTUCKY
TENNESSEE
Moss
Sugar Grove
Westmoreland
Fountain Head
LAFAYETTE
Lafayette
Hermitage Springs
Red Boiling Springs
Bethpage
GALLATIN
Castalian Springs
Hartsville
Willette
Haydenburg
JACKSON CO (1A7)
Hilham
Pleasant Shade
Difficult
Gainesboro
ARRIVING VFR AIRCRAFT SHOULD CONTACT NASHVILLE APPROACH CONTROL WITHIN 20 NM ON 128.35/360.7
Granville
LEBANON
Rome
Carthage
Elmwood
Chestnut Mound
New Middleton
Baxter
Double Springs
COOKEVILLE
Shop Springs
Watertown
ONE GRAND
UPPER CUMBERLAND REG (SRB)
See NOTAMs/Directory for Class D eff hrs
86°
VOR/DME 108.0 SYI
VORTAC 115.8 GQO
30′

Fig. 11.1

The airplane that you will be flying is a typical single-engine trainer. It has an Indicated Airspeed while cruising of 95 knots at 2,300 RPM. Use the Compass Correction Card of figure 6.3 for Deviation errors and the Airspeed Calibration table of figure 7.1. The airplane flown will have performance characteristics like those shown in figure 8.5 (Cruise) and figure 9.1 (Climb).

ALTITUDE SELECTION

To get started, select an altitude. The highest terrain or obstruction along the course is 1,800 feet, which is determined by looking at the blue number sets in the Latitude/Longitude boxes on the chart. You should not cruise lower than 1,500 to 2,000 feet above the surface. 1,500 feet AGL higher than the obstruction altitude of 1,800 MSL is 3,200 feet MSL, so the altitude selected should be at least 3,200 feet MSL or higher.

Using a Navigation Plotter, determine the trip's True Course (TC). The TC is approximately 322°. The 2°W Isogonic Line passes just East of the Upper Cumberland Airport so the Magnetic Course for the flight will be 324°. This places the MC in the Even thousand plus 500 side of the compass (180° to 359°). So the available VFR altitudes above 3,200 feet for this trip are 4,500, 6,500, and 8,500.

SPEED CALCULATIONS

Next, determine how the Ground Speed will be affected by the changing winds at these altitudes. To help determine these numbers the following table was constructed from the Winds and Temperatures Aloft Forecast:

	TC	Wind. Dir.	Wind Vel.	Tem	TAS	WCA	GS
4,500	322°	210°	17 kts	+16	101	8°L	106
6,500	322°	240°	20 kts	+13	105	11°L	100
8,500	322°	280°	26 kts	+07	108	9°L	88

The Wind and Temperature numbers for 4,500 feet had to be interpolated from off the FD report. The numbers at 6,500 came from the 6,000-foot FD report and the numbers at 8,500 came from the 9,000-foot FD.

The IAS was 95 knots, which is 92.5 rounded up to 93 knots by using the Airspeed Calibration chart (figure 7.1). The TAS was calculated at each level using a Pressure Altitude of 4,500 feet and +16°C, 6,500 feet and +13°C, and 8,500 feet and +07°C respectively. Each wind correction angle and Ground Speed are the result of three different "Wind Face" problems using the TC (322° in every case), the TAS for the individual altitudes, and the wind direction and velocity for each individual altitude. Work these problems yourself and verify the numbers!

From the table it becomes easy to see that 4,500 feet is the best altitude. This altitude is higher than 3,200 feet (the altitude that was determined to be a safe altitude to pass over terrain and obstructions). An altitude of 4,500 feet also meets the VFR cruising rules and provides the fastest Ground Speed.

Now that the cruising altitude has been determined (4,500 feet), the top line of the Navigation Record can be calculated. This line is positioned on the Navigation Record between the first checkpoint (Departure) and the second checkpoint (Top of the Climb). This line is also pointed out by the arrow labeled "A." All information on this line will pertain to the climb. It is necessary to keep the "climb" information separate from the "cruise" information because the airplane travels slower, and burns additional fuel during the climb.

The Navigation Record is designed to be completed from the left to the right. Each column of information is needed to proceed to the next column. The number to be placed in box 2A (the climb Indicated Airspeed) comes from the Climb Performance Chart, which is figure 9.1. On the Climb Performance Chart the third column from the left is the Climb IAS in knots. These numbers are the Indicated speeds that the pilot should fly when climbing, in order to obtain the airplane's maximum climb performance. You can see that climbing to 4,500 feet the pilot should fly an Indicated speed of 65 knots. This number should be placed in box 2A.

Figure 11.2 is the Completed Navigation Record for this example flight. Follow along as we go step-by-step in filling out the Navigation Record.

Now calculate the Calibrated Airspeed using the Airspeed Calibration table of figure 7.1. As you can see, 65 knots (the IAS) is not directly listed on the Calibration table, so interpolation is required. Note that 65 knots IAS is halfway between 60 and 70 knots, which are on the table. To determine the correct CAS, go halfway between the CAS numbers of 59 and 69. The answer is 64 knots. Please refer back to Chapter 7 for the discussion about why IAS and CAS are different. In this example the numbers are just one knot apart (IAS 65 and CAS 64) but do not let the small difference between the two lead you to disregard the difference.

For calculation purposes on the ground during pre-flight planning, it usually is acceptable to assume that your cruising MSL altitude is equal to the pressure altitude. There will be cases when this will not be true, such as extremely cold or hot days. Consult with your flight instructor for details. For box 4A "Pressure

The VFR Navigation Record

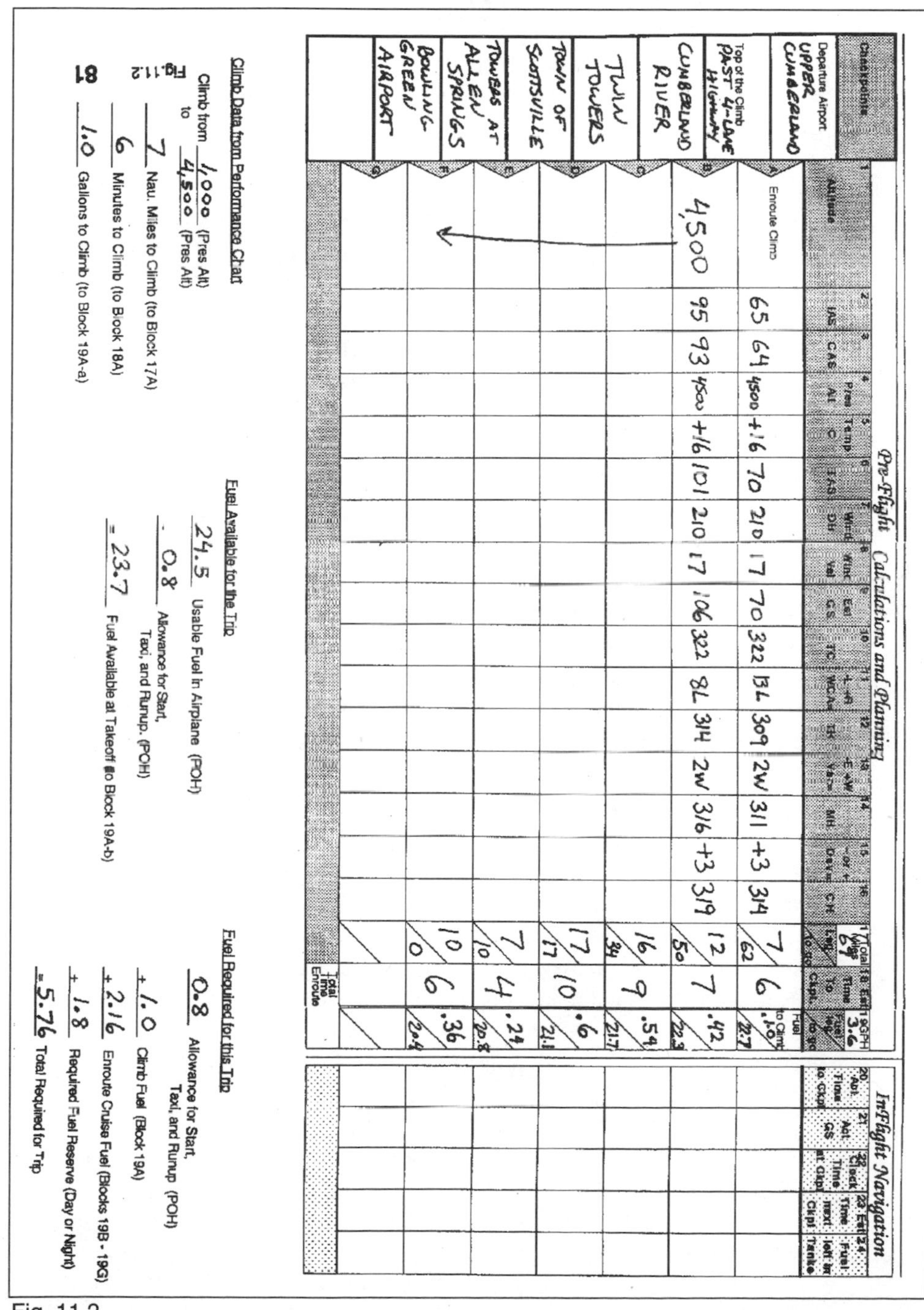

Pre-Flight Calculations and Planning / InFlight Navigation

Checkpoints		1 Altitude	2 IAS	3 CAS	4 Pres Alt	5 Temp C	6 TAS	7 Wind Dir	8 Wind Vel	9 Est GS	10 TC	11 -L +R WCA	12 TH	13 -E +W Var	14 MH	15 -or+ Dev	16 CH	17 Total Miles 69 Leg / to go	18 Est Time To Ckpt.	19 GPH 3.6 Fuel leg / to go	20 Act Time to Ckpt	21 Act GS	22 Clock Time at Ckpt	23 Est Time next Ckpt	24 Fuel left in Tanks
Departure Airport: UPPER CUMBERLAND	A	Enroute Climb	65	64	4500	+16	70	210	17	70	322	13L	309	2W	311	+3	314	7 / 62	6	Fuel to Climb 1.0 / 22.7					
Top of the Climb: PAST 4-LANE HIGHWAY	B	4,500	95	93	4500	+16	101	210	17	106	322	8L	314	2W	316	+3	319	12 / 50	7	.42 / 22.3					
CUMBERLAND RIVER	C																	16 / 34	9	.54 / 21.7					
TWIN TOWERS	D																	17 / 17	10	.6 / 21.1					
TOWN OF SCOTTSVILLE	E																	7 / 10	4	.24 / 20.8					
TOWERS AT ALLEN SPRINGS	F																	10 / 0	6	.36 / 20.4					
BOWLING GREEN AIRPORT	G																								
																			Total Time Enroute						

Climb Data from Performance Chart

Climb from 1,000 (Pres Alt) to 4,500 (Pres Alt)

7 Nau. Miles to Climb (to Block 17A)

6 Minutes to Climb (to Block 18A)

1.0 Gallons to Climb (to Block 19A-a)

Fig. 11.2 81

Fuel Available for the Trip

24.5 Usable Fuel in Airplane (POH)

- 0.8 Allowance for Start, Taxi, and Runup. (POH)

= 23.7 Fuel Available at Takeoff (to Block 19A-b)

Fuel Required for this Trip

0.8 Allowance for Start, Taxi, and Runup (POH)

+ 1.0 Climb Fuel (Block 19A)

+ 2.16 Enroute Cruise Fuel (Blocks 19B - 19G)

+ 1.8 Required Fuel Reserve (Day or Night)

= 5.76 Total Required for Trip

Fig. 11.2

Altitude" the cruise altitude of 4,500 feet is written. In actual practice the altitude on this leg is ever-changing because this is a climb leg. However, if 4,500 feet is used in calculations the answers will be the "worse case" and this is good for planning purposes. The pilot should never expect and plan using the "best case" numbers. Using "worst case" numbers, but expecting the performance to be somewhat better will automatically produce a built-in "safety factor" within the numbers.

An altitude of 4,500 feet is the Pressure Altitude and +16°C is the temperature derived from the Winds and Temperatures Aloft table. We now have the three numbers that are necessary to calculate the True Airspeed (TAS). Those numbers are CAS, Pressure Altitude, and Temperature. Using the Calculator Side of a Flight Computer, line up the Pressure Altitude of 4,500 feet across from +16°C. Then find the CAS on the wheel, and read TAS directly across from CAS on the outside scale. (Refer to Chapter 7 for the exact procedure.) This calculation yields a TAS of 70 knots.

Boxes 7A and 8A are the Wind Direction and Wind Velocity numbers that are taken from the Winds Aloft table. The Winds Aloft report does not actually predict for the altitude of 4,500 feet, but you should recall that these numbers were interpolated from the FD report. The wind direction was determined to be 210° with a velocity of 17 knots. Now all the numbers required to calculate the Ground Speed are available: the True Course, the TAS, the Wind Direction, and the Wind Velocity. Using the Wind Face of the Flight Computer you can determine using these numbers that the Ground Speed at 4,500 feet would be 74 knots during the climb.

The True Course is determined by using the plotter on the Sectional Chart. The use of the plotter was discussed in Chapter 4. The TC from Upper Cumberland to Bowling Green is 322°. The Wind Face problem used to find the Ground Speed will also produce a Wind Correction angle. This particular flight has a TC of 322° (northwest) and a wind from 210° (southwest). You can see that this flight will have a crosswind that will tend to blow the airplane off course to the northeast if a Wind Correction Angle is not used. The WCA calculated for this flight is 13° Left and this number is entered into block 11A. The TC of 322°, minus a 13° WCA will produce a True Heading (box 12A) of 309°.

VARIATION AND DEVIATION

The Variation that will affect this flight is 2° West. This information comes from the Sectional Chart. The 2°W Variation line can be found just east of the Upper Cumberland Airport and runs almost parallel to the True Course to Bowling Green. The TH of 309° is affected by the Variation. The TH of 309° plus the 2° of West Variation produces a Magnetic Heading (box 14A) of 311°.

Using the Compass Correction Card, figure 6.3, the Deviation is determined. The MH of 311° is closest to the entry for 300° on this airplane's card. This airplane has a $+3^o$ Deviation error when on the 300° MH. The $+3^o$ error should be used for this example flight and listed in box 15A. The 311° MH plus the 3° Deviation error yields a Compass Heading of 314° and this number is placed in box 16A. After considering the wind drift, Variation, and Deviation, the pilot should climb from the Upper Cumberland Airport to 4,500 feet while holding an IAS of 65 knots and a CH of 314°. This will produce the best performance and keep the airplane on course to Bowling Green.

CLIMB CONSIDERATIONS

The climb performance numbers of "Distance to Climb," "Time to Climb," and "Fuel Burned in the Climb," come from the Climb Performance Chart (figure 9.1). As discussed in Chapter 9, this chart indicates climbs from sea level. But the Upper Cumberland Airport is not located at sea level, rather it is up at 1,025 feet above sea level. For climb performance numbers we should calculate a climb from 1,000 feet to 4,500 feet. The distance to climb must be interpolated. 4,500 feet is not directly on the chart, but is halfway between 4,000 and 5,000. Using the "distances" of 8 miles and 10 miles, respectively, we see that a sea level to 4,500-foot climb would take 9 miles. Now subtract from the sea level to 1,000-foot distance interval because Upper Cumberland is already at 1,000 feet and a sea level to 1,000-foot climb is not necessary. The calculation is 9 miles minus 2 miles equals 7. The airplane will fly past 7 miles of the flight during the climb to 4,500 feet. The Top of the Climb will occur 7 miles from Upper Cumberland. Remember that this number is based on a condition of no wind. If in fact there is a strong tailwind or headwind present, this number can be adjusted. In this example flight, the wind is a crosswind so the number (7 miles) from the chart should be used, and written in the upper half of box 17A.

Using the same procedure used to produce the "Distance to Climb" determine the Time and Fuel Burn in the climb. The chart yields a "Time to Climb" of 6 minutes and a "Fuel Burn in the Climb" of 1.0 gallon. This means that the "Top of the Climb" will occur 6 minutes after takeoff and the airplane will have 1.0 less gallons than when the airplane left the ground. The Time to Climb of 6 minutes is written in box 18A, and the Fuel Burn to Climb of 1.0 is written in the upper half of box 19A. This completes the Climb Leg calculations.

CRUISE CONSIDERATIONS

The cruise leg calculations begin on the second line of the Navigation Record, labeled line "B" and will be used for all remaining lines listed "C" through "G." The Cruise Indicated Airspeed must come from the pilot's prior knowledge of how the airplane usually performs when in straight and level flight and a cruise power setting in place. For this example flight the IAS of 95 knots is used because this cruise speed is typical of light trainer airplanes. The IAS of 95 knots is written in box 2A. The CAS is determined as before using the Airspeed Calibration Chart. Doing this produces a CAS of 95 knots and this number goes into box 3B. Boxes 4B and 5B are the Pressure Altitude and Temperature and remain the same as boxes 4A and 5A. But the TAS is faster and is calculated to be 101 knots, which goes into box 6B. The Wind Direction and Velocity are the same (210° and 17 knots) in boxes 7B and 8B and the TC is the same in box 10B. But the WCA will not be the same during the cruise as it was while climbing. This is due to the fact that the airplane is traveling faster through the air while cruising and therefore the wind will not cause the airplane to drift as far. A new calculation for WCA and GS is required here. The cruise WCA is determined to be 8° Left (box 11B) rather that 13° Left as it was during the climb. Using the 8° Left WCA the cruise True Heading is 314° (box 12B). The Variation and Deviation is the same in climb and cruise, but since the TH is different in cruise the other direction numbers will also be different. The cruise MH will be 316° (box 14B) and the cruise CH will be 319° (box 16B). The airplane's final Compass Heading in cruise flight is 5° different than in climbing flight because of the difference in Ground Speed. The airplane's GS has increased to 106 knots (box 9B). At the Top of the Climb, the pilot must lower the nose of the airplane, accelerate, reduce the power, lean the engine, and adjust the heading to allow for the faster cruise speed.

Figure 11.2 has checkpoints already chosen for the flight. The Top of the Climb (TOC) will cover 7 miles. Indicate on the chart the position of the TOC. From the TOC mark the start measuring with a plotter the mileage between all other checkpoints. The distance between the TOC and the Cumberland River is 12 miles. It is 16 miles between the Cumberland River crossing and the Twin Towers that are east of the Lafayette Airport. Another 17 miles between the Towers and the town of Scottsville. Scottsville to the Towers at Allen Springs is 7 miles, and from Allen Springs to the Bowling Green Airport is 10 miles. A total cross-country distance of 69 miles from Upper Cumberland to Bowling Green. The Total Miles is written in the heading of the Navigation Record at column 17. Each individual mileage breakdown is placed in the upper half of boxes 17B, 17C, 17D, 17E, and 17F. It is always important to know how much farther the trip has yet to go when in flight. This information helps with decisions that might arise from poor weather or engine problems. The miles "To Go" information is found in the lower half of the boxes in column 17. Starting from the top of the column each "To Go" amount can be found by subtracting top to bottom. In this example there is a total of 69 miles. 69 miles minus the miles to the TOC (7 miles) leaves 62 miles yet to be flown. The next leg is 12 miles long. 62 minus 12 leaves 50 miles "To Go" when over the Cumberland River. Each leg is subtracted until zero miles are left at Bowling Green. By doing this a column of distances is listed that will be used in the next calculations.

TIME AND FUEL

Column 18 lists the Estimated Times to each checkpoint. The top line has already been determined for the climb from the Climb Performance Chart, but all cruise times must be calculated from the Flight Computer. Chapter 8 taught how to complete Time/Speed/ Distance problems. Now put those problems to practical use. The Ground Speed is known. It was calculated to be 106 (box 9B). The distance from the TOC to the Cumberland River is 12 miles. Using the Calculator Side of the Flight Computer it is determined that it takes 7 minutes to travel 12 miles when traveling at 106 knots. Rounding off to the nearest minute is acceptable. In turn, the estimated time between each leg is calculated and listed in boxes 18C, 18D, 18E, and 18F.

Now that the time between legs is known, the Fuel Burn on each leg can be determined. Refer to figure 8.3, the Fuel Consumption Table. The airplane will cruise at 4,500 feet (use the 4,000-foot numbers on the table), the temperature at cruise is +16°C, and the engine is set at 2,300 RPM. The standard temperature at 4,000 feet is 7°C. The actual temperature at 4,000 feet (calculated for this sample flight) is 16°C. The actual temperature is approximately 10° warmer than standard. The Fuel Consumption table does not have a column for 10° above standard, therefore you must split the difference between the Standard column and the 20° Above Standard column. The Fuel Burn Rate per hour or Gallons per Hour (GPH) is 3.7 at 4,000 feet with 2,300 RPM at Standard Temperature. The GPH is 3.5 at 4,000 with 2,300 RPM at 20° Above Standard. Therefore the GPH at 10° Above Standard would be 3.6. This means that for every hour that the engine operates under these conditions there will be 3.6 less gallons of fuel in the tanks. The GPH rate of 3.6 is written at the top of column 19.

The fuel burned in the climb has already been determined to be 1.0 gallons from the Climb Performance Chart, but the cruise fuel burn has yet to be determined. Using the Times for each leg in column 18, and the GPH rate, the fuel burn per leg can be determined. While flying from the TOC to the Cumberland River the engine will burn .42 gallons. This is calculated by a Flight Computer solution. How much fuel is used in 7 minutes at a rate of 3.6 GPH? The answer is .42 gallons. Each leg's fuel burn is calculated and recorded in the upper half of boxes 19B, 19C, 19D, 19E, and 19F.

Now the critical question. Will the airplane fly all the way to Bowling Green or will it run out of fuel in flight prior to reaching Bowling Green? First, we must determine how much total fuel is available for the flight. Use figure 11.3 as a table to calculate the available fuel.

Fuel Available for the Trip

________	24.5	Usable Fuel in the Airplane
minus ________	0.8	Allowance for Start, Taxi, & Runup
equals ________	23.7	Fuel Available at Takeoff (box 19A-b)

Fig.11.3

I strongly recommend that the fuel tanks be filled to capacity for any cross-country flight. There will be circumstances later in a flying career where it will be advisable to depart with less than full tanks, but those situations arise with larger and heavier airplanes. For light, single engine airplanes, the tanks should be full unless the airplane would become overweight.

For our example flight, the airplane's total usable fuel is 24.5 gallons. The airplane's Climb Performance Chart (figure 9.1) has a note at the bottom concerning the Start, Taxi, and Runup fuel allowance. The note says "Pilots should ADD 0.8 Gallons to allow for Start, Taxi, and Runup." This note is a manufacturer's estimate. If the tanks are completely full when the engine is started, they will no longer be full at takeoff after starting the engine, taxiing to the runway, and completing an engine runup check. The estimate is that at takeoff there will be 0.8 gallons less than full tanks. If 24.5 is completely full, then 23.7 gallons will be available at takeoff.

The rest of column 19 can now be completed. If the airplane takes off with 23.7 gallons, and the climb to 4,500 feet burns 1.0 gallons, then there will be 22.7 gallons remaining in the tanks at the TOC. This number is written in the lower half of box 19A. The lower half is labeled with a lower case "b."

The leg between the TOC and the Cumberland River will burn .42 gallons. That means that there will be approximately 22.3 gallons remaining. Subtracting off each leg's fuel burn produces fuel remaining in the tanks at each checkpoint until reaching Bowling Green. At Bowling Green there should be approximately 20.5 usable gallons still in the airplane. Why is it important to know how much fuel is left in the tanks at any checkpoint? If the weather ever becomes bad and the pilot is required to fly around the weather, or even fly to an unplanned airport, it will be very helpful to have an idea of how much fuel is left. This will help the pilot determine how far he or she can fly around or to another airport.

Finally, the last important question. How much fuel am I legally required to carry on this flight? Use figure 11.4.

Fuel Required for the Trip

	0.8	Allowance for Start, Taxi, and Runup
plus	1.0	Climb Fuel (from box 19A)
plus	2.16	Enroute Fuel (from boxes 19B through 19G)
plus	1.8	Required Fuel Reserve (Day or Night)
equals	5.76	Total Required for the Flight

Fig. 11.4

The Federal Aviation Regulations make it illegal to land an airplane returning from a flight away from an airport to have less that 30 minutes of fuel during the day or 45 minutes of fuel at night. The amount of fuel in gallons that constitutes a 30 or 45 minute fuel reserve can be calculated on the flight computer. Using the cruise fuel consumption of 3.6 GPH, a 30 minute day fuel reserve would be 1.8 gallons. A 45 minute night fuel reserve would be 2.7 gallons.

The sample flight from Upper Cumberland to Bowling Green, if flown during daylight would legally require 5.76 gallons. This number was determined using figure 11.4.

0.80	Allowance for Start, Taxi, and Runup
1.00	Climb Fuel from box 19A
2.16	Enroute Fuel from Boxes 19B, 19C, 19D, 19E, and 19F
1.80	Daylight Fuel reserve at 3.6 GPH
5.76	Total Fuel Required for the Flight

Now compare the Fuel Available at Takeoff, which is the final answer to "Fuel Available for the Trip" (figure 11.3), with the final answer to "Fuel Required for the Trip" (figure 11.4). If the available fuel is greater than the required fuel, then the flight can be completed non-stop. But, if the available fuel is ever less than the required fuel, then an intermediate fuel stop must be planned enroute!

This flight example has been over a short distance. The numbers have been small. The speeds have been slow. The fuel burned has been little. But do not let the small numbers fool you. It is important to be able to calculate these numbers accurately. If you plan to pursue a professional flying career, you will eventually fly bigger airplanes that will have faster speeds, higher climbs, and greater fuel burns. By understanding the smaller numbers now, you will have a smoother transition to the larger numbers later. The fuel required may only be 5.76 gallons today, but it may become 576 gallons in your future.

As a practice exercise, use a blank Navigation Record and plan the return trip from Bowling Green to Upper Cumberland. Use all the same performance charts, winds, temperatures, and power settings. Good Luck!

12
The VFR Flight Plan

THE FEDERAL AVIATION ADMINISTRATION HAS DEVELOPED A SAFETY system to aid VFR pilots on cross-country flights: The VFR Flight Plan. The Flight Plan, as the name implies, is a record of a pilot's intentions. A pilot "files" the Flight Plan with the nearest Flight Service Station prior to departure. This can be accomplished over the telephone, in person, or over a computer. Once the Flight Plan is on file, the pilot will "activate" the Flight Plan by calling the FSS on the radio. Once activated, the Flight Plan is passed over to the FSS that is nearest to the trip's destination. When the pilot arrives at the destination, he or she calls the FSS and "cancels" the Flight Plan.

FREE INSURANCE

The Flight Plan is actually an insurance policy that costs nothing. Hopefully the pilot will arrive at the destination safely and will cancel the Flight Plan. But what if the pilot runs into trouble along the way? What if the pilot must make an emergency landing? The Flight Plan has an Estimated Time Enroute section. Adding this enroute time to the time when the Flight Plan is activated gives a proposed time of arrival at the destination. The pilot should call to cancel the Flight Plan within thirty minutes of this proposed arrival time. If after thirty minutes the pilot does not call to cancel, the FSS will begin a telephone and radio search. The FSS will first call the destination airport to determine if the airplane is safely on the ground. If it is the middle of the night and nobody is on duty at the airport, the FSS may call the local sheriff or police. The police go to the airport and look around for the airplane. They can identify the airplane by its tail number. If the airplane cannot be found at the destination airport, the FSS will call airports that are along the pilot's proposed route of flight. They will call other Flight Service Stations or approach controls, or anybody who may have had a radio contact with the airplane. After all this fails to find the airplane, an actual search is begun. The people who actually do the search will vary from place to place. Sometimes the search is conducted by the National Guard, Coast Guard, Highway Patrol, Civil Air Patrol, or other agency.

Imagine that you are the pilot of an airplane that experienced engine problems. You had to make a forced landing. You select an area that looks to be the best place for an emergency landing. The touchdown is made but the airplane's nose wheel falls in a hole and the airplane is tossed on its back. During the sudden stop you are thrown forward and hit your head. You are momentarily knocked out. When you become aware again, you have a terrible pain in your back and you are bleeding from one arm. You are hanging upside down in the seat belt. The nearest telephone is probably miles away, and you are not sure you can climb from the plane with a back injury. Now is the time that you will be very glad that you had filed a VFR Flight Plan.

VFR Flight Plans are actually optional. Unlike IFR Flight Plans, they are not required for flight. But if you do not file a Flight Plan and have an accident, help will not be on the way. There have been cases where pilots have actually died because they did not get prompt medical care following an accident. Having a Flight Plan filed means that help is on the way.

Sometime ask an FSS employee "What is the biggest problem with VFR Flight Plans?" They probably will tell you the biggest problem arises when a pilot on a Flight Plan arrives safely at the destination, but simply forgets to call and cancel. Ninety-nine percent of the overdue pilots have just not remembered to call. The FSS usually finds these pilots drinking a soft drink in the pilot's lounge when they call. I asked an FSS briefer once this question: What if a pilot lands at an unattended airport at night, then pushes the airplane into a hangar and locks the doors? He forgets to cancel his Flight Plan, but drives to a local restaurant for dinner. There is nobody at the airport to take a phone call. The airplane will not be discovered by police because it's in a locked hangar. The pilot is not at the telephone number written on the Flight Plan. The FSS cannot find the airplane so eventually a search begins. Airplanes are dispatched to fly the pilot's route while listening for a distress signal. Many planes, worker-hours, and money is spent. What will you do to that pilot when you finally get to him? The FSS briefer surprised me. He said: We won't do anything to him. We will let him know what trouble he caused, but no enforcement action will be taken. If we did penalize this pilot the chances are that he would never again file a Flight Plan, and this guy is the very person who is likely to need it the most!

The moral of the story is to file VFR Flight Plans for all trips and remember to cancel. Many airports have signs posted that read "Have you closed your Flight Plan?" I know of one airport that has such a sign placed just above the urinal in the men's room. They figure this is where the pilot is likely to go first after landing. (I do not know if such a sign was in the women's room or where it would be placed. My research did not go that far!)

THE FAA FLIGHT PLAN FORM

The Flight Plan form itself (figure 12.1) is actually three forms in one. **Block 1** of the Flight Plan is labeled "Type." The pilot will check the box that applies to the flight. The selections are: VFR, IFR, or DVFR. The purpose of a VFR Flight Plan has been discussed. The VFR Plan is used in good weather. The IFR Flight Plan can be used during good weather, in reduced visibility or even in the clouds. To use an IFR Flight Plan the pilot must be Instrument Rated and instrument current. The DVFR Flight Plan is a "Defense" Plan and is used when flying into the airspace of the United States from outside the country's boundaries.

Block 2 is the airplane's full tail number, starting with the letter "N" for all United States registered airplanes. In **Block 3** is listed the type of airplane that will be flown. Abbreviations such as C-152 or PA-44-180 are acceptable. Also in Block 3 is a code for the "special equipment" that is on board the airplane. A letter is used to represent equipment such as a transponder, DME, RNAV, Altitude Encoding (Mode C) or the lack of equipment such as, No Transponder. In this way space is saved and neither the pilot nor the FSS briefer has to jot down all equipment. The current codes as of this writing are:

- A DME and Transponder with Mode C
- B DME and Transponder without Mode C
- C RNAV, Transponder but without Mode C
- D DME but no Transponder
- R RNAV and a Mode C Transponder
- T Transponder without Mode C
- U Transponder with Mode C
- W RNAV but no Transponder
- X No Transponder

Please consult the most recent copy of the Aeronautical Information Manual to verify these codes. They have changed from time to time. As new navigation systems become common more codes will be added.

U.S. DEPARTMENT OF TRANSPORTATION
FEDERAL AVIATION ADMINISTRATION

FLIGHT PLAN

(FAA USE ONLY) ☐ PILOT BRIEFING ☐ VNR ☐ STOPOVER | TIME STARTED | SPECIALIST INITIALS

1. TYPE	2. AIRCRAFT IDENTIFICATION	3. AIRCRAFT TYPE/ SPECIAL EQUIPMENT	4. TRUE AIRSPEED	5. DEPARTURE POINT	6. DEPARTURE TIME		7. CRUISING ALTITUDE
					PROPOSED (Z)	ACTUAL (Z)	
☐ VFR							
☐ IFR							
☐ DVFR			KTS				

8. ROUTE OF FLIGHT

9. DESTINATION (Name of airport and city)	10. EST. TIME ENROUTE		11. REMARKS
	HOURS	MINUTES	

12. FUEL ON BOARD		13. ALTERNATE AIRPORT(S)	14. PILOT'S NAME, ADDRESS & TELEPHONE NUMBER & AIRCRAFT HOME BASE	15. NUMBER ABOARD
HOURS	MINUTES		17. DESTINATION CONTACT/TELEPHONE (OPTIONAL)	

16. COLOR OF AIRCRAFT

CIVIL AIRCRAFT PILOTS. FAR Part 91 requires you file an IFR flight plan to operate under instrument flight rules in controlled airspace. Failure to file could result in a civil penalty not to exceed $1,000 for each violation (Section 901 of the Federal Aviation Act of 1958, as amended). Filing of a VFR flight plan is recommended as a good operating practice. See also Part 99 for requirements concerning DVFR flight plans.

FAA Form 7233-1 (8-82) CLOSE VFR FLIGHT PLAN WITH ______________ FSS ON ARRIVAL

Fig.12.1

The Special Equipment code or "suffix" is written in **Block 3** after the airplane type. C-172/U or BE76/R or even J3 Cub/X.

Block 4 is where the True Airspeed is listed in knots. It is assumed that this is the TAS in the cruise configuration. This number then can be taken directly from the Navigation Record, specifically box 6B.

Block 5 is the Departure airport and **Block 9** is the Destination airport. It is best that the airport's three letter identification be used here. Virtually all airports now have identifiers that are unique to that particular airport and are unmistakable.

Block 6 is the departure time given in "Zulu" time. The Zulu time is also referred to as the Universal Time Conversion (UTC), Greenwich Civil Time (GCT) and Greenwich Mean Time (GMT). It is necessary to standardize time so that the exact takeoff time can be anticipated. If a pilot says that the takeoff is proposed to be 7 o'clock, does this mean 7 in the morning or 7 in the evening? Does this mean 7 o'clock Central Time or Pacific Time? Does this mean Standard or Daylight Savings Time?

This is always very confusing for pilots, but the fact is that the Earth is round and is turning. The Sun's rays come only from one direction; therefore, the Sun can be overhead at only one location at a time. People have their lives and daily activities matched to the Sun's location throughout the day. We want to get up when the Sun is in the East, have lunch when the Sun is overhead, and sleep after the Sun has gone over the western horizon. Furthermore, people want to maintain this schedule regardless of where on the Earth they live. This means that somewhere on the Earth it is lunch time because the Sun is overhead and simultaneously it is the middle of the night in another location. To stay on schedule, regardless of your Earthly location, time zones were invented. There are a total of 24 time zones on Earth, each approximately a 15 degree slice of the Earth.

The Sun is stationary relative to the Earth. The Earth turns in front of the Sun. When a particular location on Earth turns under the position directly across from the Sun, it is noon at that location. Thinking of it this way, noon and all other times are stationary, the locations on Earth simply rotate from position to position, which is the same thing as saying locations rotate from time to time.

But back on Earth, air traffic controllers need to know when to expect airplanes to take off and arrive at destinations. Flight Service Station workers need to know when to anticipate activating Flight Plans. Pilots must calculate time in the air regardless of how many time zones are crossed. Therefore the Zulu time conversion was established. With a standardized system of time we can all mean the same thing regardless of what the local time may be. The system of standardized time that is in place aligns itself with the same starting point as the lines of Longitude. Review Chapter 3 and recall the British Royal Observatory and its role in determining the Prime Meridian. Zulu Time also has a starting point and it is also at the Prime Meridian in Greenwich, England. This system was designed in 1884. Because the system was designed in England, American pilots must convert local times to Greenwich, England time to arrive at the international standard. The following conversions should be used:

To convert from:	To Universal Time Conversion (UTC)
Eastern Standard Time	Add 5 hours
Eastern Daylight Time	Add 4 hours
Central Standard Time	Add 6 hours
Central Daylight Time	Add 5 hours
Mountain Standard Time	Add 7 hours
Mountain Daylight Time	Add 6 hours
Pacific Standard Time	Add 8 hours
Pacific Daylight Time	Add 7 hours
Alaska Standard Time	Add 9 hours
Alaska Daylight Time	Add 8 hours
Hawaii-Aleutian Standard Time	Add 10 hours
Hawaii-Aleutian Daylight Time	Add 9 hours

Block 6 is divided into two parts labeled "proposed" and "actual." The pilot should estimate the time of departure and place this time in the "proposed" box. Then later, just before takeoff, the pilot should jot down the true departure time in the "actual" box.

After determining the best altitude to fly (refer to Chapter 10) the pilot lists the cruising altitude in **Block 7**. One of the most important blocks is **Block 8**. In this block the pilot lists the Route of Flight. Most VFR flights are planned to go "direct." In other words, the flight path between the departure airport and the destination airport is a straight line. When this is the case the word "direct" is all that is needed here. But if the flight path is not straight, the exact course should be indicated. This is important because if a search is ever actually conducted, the searchers will follow the course described in Block 8. If the path flown and the path listed in Block 8 were different, you would send the searchers along the wrong path and this would lengthen your time of rescue.

Block 10 is the Estimated Time Enroute. This information can be taken directly from the Navigation Record (bottom of column 18). The FSS will add the Block 10 time to the clock time that the Flight Plan is activated. This determines when the airplane should arrive at destination.

Block 11 and **Block 13** have limited use on VFR Flight Plans. The "Remarks" block is the location where "notes" or even "requests" are left for the FSS or Air Traffic Controllers. Usually this pertains to IFR Flight Plans. The "Alternate Airport(s)" block means different things depending on whether the Flight Plan is VFR or IFR. For VFR Flight Plans one or more airports can be listed, usually airports that are along the way to the destination airport. By listing an airport in Block 13 of a VFR Plan, the pilot is simply planning for the possibility that an unscheduled landing might be needed. The pilot can select any airport for this block. Block 13 of an IFR Plan is different. There can only be a single IFR alternate and many airports cannot be used, depending on the facilities and forecast weather.

Block 12 lists the amount of fuel in the airplane. Take special notice of the fact that the fuel level is not given in gallons on a Flight Plan. The fuel must be calculated in Hours and Minutes. During pre-flight planning the pilot must allow for start, taxi, runup, and climb, then use the cruise fuel consumption to determine how long the engine would run until dry. If a search is ever begun, the fuel on board determines the limit of the search area. Also, the searchers know that after the fuel on board time has elapsed from the takeoff time, the airplane is on the ground somewhere.

Block 14 is where the pilot's name, address, telephone number, and home base is listed. This information would start the search if the flight becomes overdue. The phone number given should be to a location where someone will be present to answer the phone. Do not use a flight school, airport manager's office, or FBO if these locations will be closed during the flight.

Block 15 and **Block 16** can be a little scary. In truth the information in these two blocks is there just to help searchers locate a airplane that has been in an accident by its color, and to let the search team know how many people to look for.

Block 17 although listed as "optional" is a good idea. It is the person and or telephone number of the location to which you are traveling. If you flew to a vacation spot, but became overdue, it would do no good for FSS to call your home phone, because you will not be there! Leave a destination contact, a friend, hotel, ski instructor, anything.

Then at the bottom of the Flight Plan form are the words "Close VFR Flight Plan with _________ FSS on Arrival." You should fill in this blank with the FSS you will call to cancel. Writing it down will help you remember to do it.

ACTIVATING THE FLIGHT PLAN

When calling in a Flight Plan the pilot should ask the FSS what radio frequency is best to use when activating the plan. The VFR Flight Plan does not go into force until the pilot is in the air and calls via radio to say: "Smithtown Radio, this is Cessna 1234A. Please activate my VFR Flight Plan to Jonesville." The pilot must therefore be able to contact an FSS while in the air. (More on in-flight contact with FSS in Chapter 14.)

Often the pilot will be unable to activate the plan immediately after takeoff. The pilot may have to climb to reach an altitude high enough to contact an FSS. Other times when departing from an airport with Radar control, it may be several minutes before it is possible to switch from the Departure control frequency to the FSS frequency. The pilot must ask the Departure controller for permission to leave the frequency when being Radar vectored. Pilots should use good judgment and awareness here. Do not ask to leave the ATC frequency when headings and traffic information are being given or are likely. Just prior to takeoff the pilot should record the actual time off on the Navigation Record and the Flight Plan. When the opportunity arises for the pilot to contact the FSS the pilot should not activate the plan based on the time of the radio contact, but rather the actual time of takeoff. The pilot may be in the air several minutes before FSS contact is possible or allowed by ATC. When activating the plan after a delay the pilot should say: "Smithtown Radio, this is Cessna 1234A. Please activate my Flight Plan to Jonesville at 25 minutes past the hour" (the time of takeoff).

Remember to *PLAN* the flight accurately, *FILE* a VFR Flight Plan, *FLY* your course, and *CANCEL* when safely at the destination.

Do not ask the Control Tower or Departure controller to activate your VFR Flight Plan. These controllers are usually too busy to make a telephone call to the FSS for you. There are certain locations where a controller might offer to activate the plan for you, but pilots should consider this a courtesy, not a regular practice.

Remember to plan the flight accurately, file a VFR Flight Plan, fly your course, and cancel when safely at the destination.

13
Navigation Record in-flight and the *Missing Wind* problem

IN CHAPTER 11 THE STATEMENT WAS MADE THAT THE NAVIGATION Record is not completed until the flight has been completed. This is true because the pilot should use the Navigation Record during the flight to keep up with position and fuel consumption. I have seen pilots do a wonderful job of calculating and completing the Navigation Record, but then toss it in their flight bag and fly without consulting it. Used properly while in flight, the Navigation Record becomes the pilot's first line of defense against getting lost.

Look back to figure 8.1, the Navigation Record. Columns 20 through 24 of the Record are under the heading *InFlight Navigation.* These columns are placed together at the right side of the Record to be used while in the air. Together the information in these columns will insure that the flight is going along as predicted.

Column 18 is the *before flight* prediction of how long it will take the airplane to arrive at the Top of the Climb and the other checkpoints. Column 20 is the *Actual Time* the airplane took to reach the checkpoints. The pilot has not completed the pre-flight inspection until he or she has a reliable clock or watch and a pencil on board the airplane. The pilot should have a clipboard that can be placed across the lap so that the cockpit turns into a small office. The clipboard should have the Sectional Chart (properly folded to the area of the flight), the Flight Plan, and the Navigation Record attached. Just prior to takeoff the pilot should write in the Actual time of departure on the Flight Plan (Block 6). Then as the airplane reaches the Top of the Climb the pilot should see how much time has elapsed since the takeoff and write this elapsed time in box 20A of the Navigation Record. The Actual clock time at the TOC should also be written in box 22A. The Actual time to the TOC should then be compared to the Estimated time (box 18A). If the numbers are the same or very close this confirms that the Climb Performance chart was accurate.

After the airplane has reached the Top of the Climb and is flying level the pilot can predict the passage of the next checkpoint. The clock time at the TOC was written in box 22A. Add the time in 22A to the next leg's estimate in box 18B. Place the answer in box 23A. This tells the pilot the clock time that the next checkpoint should be below the airplane. This information is very important to compare with the time it actually takes to reach the next checkpoint. If the time predicted to reach the next checkpoint is 2:33 pm but the checkpoint is not actually crossed until 2:41 pm, then the pilot has a problem. This might mean that the Winds Aloft Forecast that was given from the FSS and used to calculate the Navigation Record is inaccurate. The airplane's groundspeed is obviously not as fast as predicted. It may be that the airplane has a stronger headwind than predicted. Or maybe the airplane has a weaker tailwind than predicted. If the groundspeed is slower the trip will take longer. This means the trip will require more fuel than predicted. Will the pilot have enough fuel, given the slower groundspeed, to make it to the destination or will an unplanned fuel stop be necessary? The pilot must decide this while in flight. You can see that a pilot who does not keep track of the flight's progress by timing and using the Navigation Record might be unaware that a problem exists. This is how pilots run out of gas in flight. They do not see what is happening around them until it is too late!

When arriving late over a checkpoint the ground speed will need to be recalculated. To find the Actual Groundspeed (box 21) a flight computer problem is needed. To find Actual Groundspeed the pilot must first know the distance and time. The distance between checkpoints is found in column 17 and the Actual Time to the Checkpoint is column 20. The pilot uses the Calculator side of the Flight Computer. He or she lines up the distance across from time on the computer wheel and reads the Groundspeed across from the pyramid. The Actual Groundspeed is then recorded in column 21. The Actual Groundspeed (column 21) should then be compared with the Estimated Groundspeed (column 9). If there is a significant difference then the fuel must also be recalculated.

In Chapter 11 a cross-country flight from Upper Cumberland to Bowling Green was planned. Now let's fly that trip and keep up with the In-Flight Navigation section of the Navigation Record. Refer to figure 11.2 for the Pre-Flight Calculations and Planning section.

IN-FLIGHT NAVIGATION EXAMPLE

The pilot, we will call her Dorothy, has completed the Navigation Record (figure 13-1) for her proposed flight from Upper Cumberland to Bowling Green. She has made all the pre-flight inspections of the airplane, and prepared the cockpit for this cross-country flight. She takes off and records the time at 10:15 am. After takeoff she turns on course and continues climbing to 4,500 feet. During the climb she contacts the Nashville FSS and activates her VFR Flight Plan to Bowling Green. When she arrives at the Top of the Climb, she pushes the nose over to the horizon, allows the airspeed to increase, reduces the power, and properly leans the mixture. After the airplane has been taken care of she looks at her watch and reads 10:22. She writes down this time in box 22A and sees that it has taken 7 minutes to reach 4,500 feet. She writes down 7 minutes in box 20A. She compares this box with box 18A. The estimate from the airplane's climb performance data was 6 minutes, so the actual climb took a minute longer. Seven miles in 7 minutes is 60 knots groundspeed.

Dorothy looks at box 18B next. This box tells how long it was predicted to go from the TOC to the next checkpoint which is the Cumberland River. Box 18B says that it should take 7 minutes to reach the river. She adds 10:22 and 7 minutes to get the Estimated Clock Time at the next Checkpoint (box 23B) of 10:29. Dorothy holds the cruise Compass Heading tightly and looks for the river.

At 10:29 the airplane has not yet reached the river as expected, but the river can be seen ahead. It is actually 10:31 when the river is passed. Dorothy writes down 10:31 in box 22B and 9 minutes in box 20B. The airplane was 2 minutes late over the river. Using the Flight Computer she quickly takes the distance between the TOC and the river (box 17) which is 12 miles and the real time it took to fly those 12 miles which was 9 minutes. This yields an Actual Groundspeed of 80 knots. Eighty knots is a significant difference from the Estimated Groundspeed that was calculated on the ground to be 106 (block 9B). What is wrong? It could be several factors. The factor that must be considered strongly is the fact that the Winds and Temperatures Aloft Forecast that was used to calculate the Estimated Groundspeed is in error. The Winds and Temperatures Aloft Forecast comes out every 6 hours. It is possible that when Dorothy received the forecast is was already 5 or 6 hours old. The wind can and will change both its speed and direction in a 6-hour time period. So the pilot must understand that an in-flight re-evaluation of the Ground Speed and Wind Correction Angle may be necessary on each trip.

Dorothy decides that another leg should be flown and evaluated before changing any total trip estimates. The leg between the Cumberland River and the Twin Towers near the Lafayette Airport is 16 miles and has been estimated to take 9 minutes. Block 22B has the actual time over the river, 10:31, so Dorothy adds 9 minutes to this and she gets a Clock Time at next Checkpoint of 10:40. She lists 10:40 in Block 23C and begins looking for the towers.

It is not until 10:43 that the airplane passes alongside the towers. Dorothy lists the Clock Time at the Checkpoint as 10:43 in Block 22C. This means that it has actually taken 12 minutes, not the estimated 9 minutes, to get to this checkpoint. Dorothy lists 12 minutes in Block 20C. A trend has developed. The airplane has been late to every checkpoint. Dorothy calculates the Actual Groundspeed for the leg. It took 12 minutes to fly the 16 miles between the river and the towers. This is a Ground Speed of 80 knots. Eighty knots is recorded in Block 21C. The Actual Groundspeed calculation for the first two legs has been 80 knots (Blocks 21B and 21C). Dorothy now believes that the Estimated Groundspeed calculation of 106 is in error. She determines to calculate the remainder of the flight using the 80-knot Actual Groundspeed number. This will mean that the flight will take longer and will burn more fuel than planned. The remainder of the flight is 34 miles. Thirty-four miles at 80 knots will take 26 more minutes. Twenty-six minutes at 3.6 GPH will require approximately 1.6 gallons. The remainder of the flight was originally calculated at 106 knots and that would have taken 19 minutes and burned 1.1 gallons. This means that the unexpected reduction in groundspeed will cost this flight an additional 7 minutes and 0.5 gallons of fuel. Dorothy knows that although inconvenient, this delay is not dangerous. The trip can still be made easily non-stop without dipping too deeply into the airplane's fuel reserve.

This flight was short, only 69 miles. Consider how this groundspeed reduction would have affected a longer flight. Pilots must eventually fly a VFR cross country that has one leg of greater than 250 nautical miles. This is one of the Commercial Pilot Aeronautical Experience requirements. What if a pilot flew a 250-mile leg with an estimated groundspeed of 106? It would take approximately 2 hours and 25 minutes. If the cruise fuel burn were still 3.6 GPH this leg would require 8.5 gallons. But if the actual groundspeed were only 80 knots as in our example problem the time would actually be 3 hours and 10 minutes. The fuel burn would become 11.5 gallons. You can see that the error between estimate and actual numbers becomes greater with distance. It is also easy to see that if you take this to the extreme, where the groundspeed is much slower and the distance is much greater, the airplane could run out of fuel while in flight.

The only way to know if the airplane is proceeding as planned is to keep track

InFlight Navigation

Checkpoints	Leg	17 Total Miles 69 — Leg / to go	18 Est Time To Ckpt.	19 GPH 3.6 — Fuel leg / to go	20 Act Time to Ckpt	21 Act GS	22 Clock Time at Ckpt	23 Est Time next Ckpt	24 Fuel left in Tanks
Departure Airport UPPER CUMBERLAND AIRPORT				Fuel to Climb					
	A	7 / 62	6	a 1.0 / 22.7b	7	60	10:22	↓	22.7
Top of the Climb PAST 4-LANE HIGHWAY									
	B	12 / 50	7	.42 / 22.3	9	80	10:31	10:29	22.2
CUMBERLAND RIVER									
	C	16 / 34	9	.54 / 21.7	12	80	10:43	10:40	21.5
TWIN TOWERS									

Fig.13.1

of checkpoints and timing between checkpoints. The worst case situation is when a pilot flies a trip but is not "navigating" the trip. The pilot is not calculating groundspeeds and making sure there is enough fuel. As the trip continues the flight becomes more and more off schedule, but the pilot is not aware that he is behind because he is not keeping track of time. Someday, somewhere, this pilot will be shocked when the engine sputters into silence. Maybe then he will wake up!

THE 60 TO 1 RATIO

The wind can also drift a pilot sideways off the course. The Groundspeed might be as predicted, but the course alignment may be off. The distance off-course will increase as the flight progresses. At 60 miles of flight the airplane will be off-course 1 mile for every 1 degree of wind drift. This is referred to as the **Rule of Sixty**. Following this ratio if a pilot flew 60 miles without correcting for a 10-degree crosswind drift, the airplane would be 10 miles off target. You can see that a small wind drift can lead to larger off-course and "lost" problems if not recognized and corrected early.

The 60 to 1 ratio can work another way. For every 60 knots of True Airspeed, one degree of Wind Correction Angle equals one degree of crosswind component. Using the ratio an airplane traveling with a True Airspeed of 120 knots, and holding a 10-degree Wind Correction angle is being subjected to a 20-knot crosswind component (60 knots x 2 = 120 so 10 degrees x 2 = 20 knots crosswind). Using the 60 to 1 ratio, pilots can visualize the condition of the airplane moving through the wind and anticipate off-course problems.

If the Winds and Temperatures Aloft Forecast can be wrong and therefore cause pre-flight planning to be wrong, how do you get it right? The pilot does have the ability to calculate the actual winds in flight so that an estimate does not have to be totally relied on. The pilot can work a Missing Wind Problem!

THE MISSING WIND PROBLEM

To determine the Actual wind while in flight the pilot must know five items. The pilot must know the True Course, the True Airspeed, the Actual Groundspeed, the Wind Correction Error, and the True Heading. All five of these items can be discovered in flight.

The True Course and True Airspeed is known even before the flight during the pre-flight planning and is recorded on the Navigation Record under columns 6 and 10. The Actual Groundspeed is calculated in flight by taking the actual time between two checkpoints when the distance between the checkpoints is known. The last two items, the actual WCA and the TH, are determined by actual trial.

To determine the actual Wind Correction angle the pilot must be able to see a checkpoint up ahead. This is where it pays to select good checkpoints. During the sample trip from Upper Cumberland to Bowling Green the TC was 322°. Let's say that on the leg between the Cumberland River and the Twin Towers the wind was causing the airplane to drift off course. The pilot looks ahead and while still about 6 to 8 miles away spots the towers. The pilot notices that the airplane's ground track is leading away from the towers. In other words, if the present heading is maintained the wind will blow the airplane to one side and the airplane will not cross over the towers. The WCA for this leg, calculated on the ground, was 8° Left, but obviously this angle is not enough to hold the proper path to the towers. The pilot turns the airplane into the wind more to a heading of 305°. This turn increases the Wind Correction angle. After a few minutes the airplane is established in a "crab" angle that will cause the airplane to cross over the towers. The pilot reads the Heading Indicator. The pilot determines that 305° is the direction that is keeping the airplane's ground track toward the towers. The proper True Heading is therefore 305°. The Wind Correction angle is the difference between the TC and TH. In this case the TC is 322° and the TH is 305°. This makes the WCA 17° Left.

In the previous discussion of the sample cross country it was determined that the leg between the river and the towers was 16 miles and took 12 minutes to fly. An Actual Groundspeed was calculated to be 80 knots. With this information the pilot has all that is needed to find the "missing wind."

The TC is 322°

The TAS is 101 knots

The GS is 80 knots

The WCA is 17° Left

Follow these steps with the Wind Face side of a Flight Computer to determine the actual wind direction and velocity.

Step 1. Place the TC under the True Index of the Flight Computer (322°).

Step 2. Slide the movable card so that the GS (80 kts) is under the center hole.

Step 3. Mark a "wind dot" at the intersection of WCA (17°L) and the TAS (101 kts).

Step 4. Turn the wheel so that the "wind dot" is under the True Index.

Step 5. The Actual Wind Direction is read under the True Index (262°). The Actual Wind Velocity is read between the "wind dot" and the center hole (33 knots).

The Winds and Temperature Aloft Forecast had predicted winds to be 210° at 17 knots, but the in-flight calculations revealed the forecast to be in error. The Actual winds were 260° at 33 knots. This difference between estimated and actual winds caused the entire trip to be changed. Pilots should always be ready to re-plan their Navigation numbers while in flight, because the atmosphere we fly in is ever changing. The pilot must be ready to change with it.

Verify these Missing Wind problems:

	TC	TAS	GS	TH	WCA	Wind Dir	Wind Vel (kts)
1.	307°	80	107	314°	**7°R**	**110°**	**29**
2.	033°	95	123	025°	**8°L**	**236°**	**32**
3.	171°	91	79	168°	**3°L**	**149°**	**13**
4.	316°	87	74	326°	**10°R**	**008°**	**19**
5.	274°	153	164	278°	**4°R**	**050°**	**16**

6. You are flying a VFR cross country flight. You fly over checkpoint "A" at 1723 zulu time and then over checkpoint "B" at 1732 zulu. Your TAS is 105 knots. Your TC is 210°, but you are flying a heading of 220° in order to remain on course. The distance from checkpoint "A" to "B" is 14 nautical miles. What is the wind's velocity and direction?
   270° at 21 knots

7. You are flying a VFR cross-country flight. You fly over checkpoint "C" at 2255 zulu time and then over checkpoint "D" at 2307 zulu. The distance from checkpoint "A" to "B" is 21 nautical miles. Your TAS is 90 knots. Your TC is 075°. The wind is causing you to fly a heading of 060° to remain on course. What is the wind's velocity and direction?
   308° at 39 knots

14
Weather briefings and in-flight weather information

PRE-FLIGHT

THE WEATHER INFORMATION THAT IS USED IN THE NAVIGATION Record must come from somewhere. In Chapter 10 we looked at the Winds and Temperatures Aloft Forecast and how to read it from a DUAT computer. In this chapter we want a more comprehensive look at the weather system and how weather information reaches the pilot. Figure 14.1 is a flow chart of how weather information works its way through the system to the pilot.

Weather Observers from across the country make hourly observations and pass their report on to the National Weather Service computer. Other observers include: Radar, Upper Air (weather balloons) and Satellite observations. The National computer acts as a clearing house for weather information. Once the information reaches the National computer, it can be accessed by many people. The weather briefer at a Flight Service Station (FSS) gets the information, as well as anyone with a computer and the ability to use the Direct User Access Terminal (DUAT). A DUAT computer gets its information from a "vendor." A vendor is a computer company that passes on weather information from the National computer to the pilots for a fee. Fortunately, as of the time of this writing, the FAA is paying the fee. Pilots can receive weather information from the FSS or from a

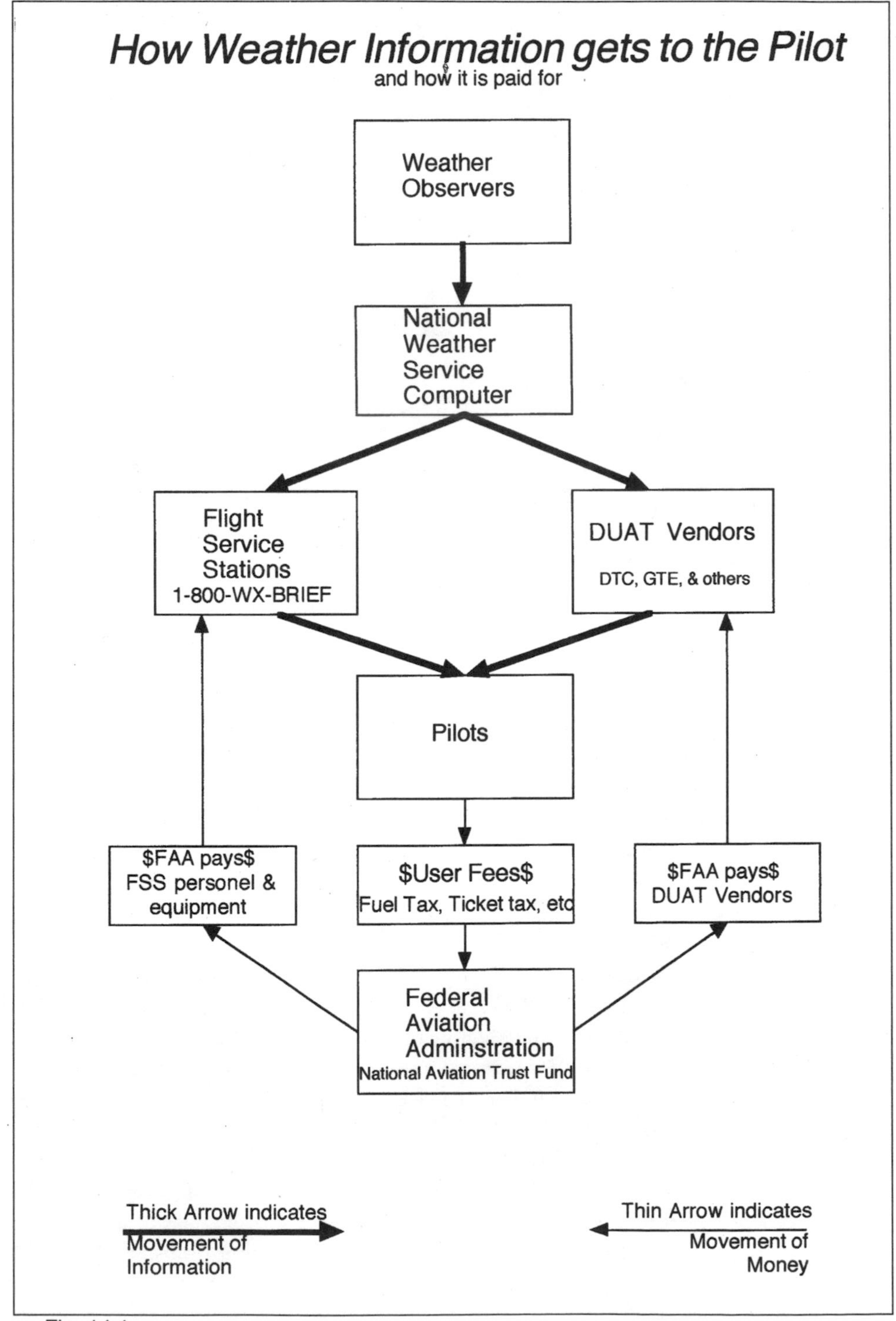
How Weather Information gets to the Pilot
and how it is paid for
Weather Observers
National Weather Service Computer
Flight Service Stations
1-800-WX-BRIEF
DUAT Vendors
DTC, GTE, & others
Pilots
$FAA pays$
FSS personel &
equipment
$User Fees$
Fuel Tax, Ticket tax, etc
$FAA pays$
DUAT Vendors
Federal Aviation Adminstration
National Aviation Trust Fund
Thick Arrow indicates Movement of Information
Thin Arrow indicates Movement of Money

Fig. 14.1

DUAT vendor. Currently, the money to pay FSS personnel and DUAT vendors comes from pilots, although the payment is made indirectly. Pilots pay into the National Aviation Trust Fund when they purchase aviation fuel. A tax for this fund is placed on each gallon of fuel sold. The Aviation Trust Fund is later tapped to pay for equipment and services provided by the FAA including FSS and DUAT service. So pilots pay their own way. But many pilots do not know how to get the most for their money. How do pilots get the best weather information so that they will have the best ability to plan their flights and decide whether or not it is safe to fly?

Today the national FSS telephone number is 1-800-WX-BRIEF. If you pick up any telephone and dial this number an FSS somewhere will answer. Usually you are first greeted by a recording and a menu of services. If you desire to speak with an FSS Briefer directly, you follow the recorded directions and remain patient. The briefer is very busy. They may keep you on hold for several minutes. When it is your turn you must be prepared. You must get the information that you need without wasting the briefer's time.

Before the FSS will give you weather information, they must first get some information from you. It is good form to volunteer the necessary information at the beginning of your conversation. The FSS needs to know 1) your full airplane number, 2) whether your proposed flight will be VFR or IFR, 3) where you are and where you are going, and 4) the time the flight will take place. After this information has been passed to the FSS, the briefer can do a better job of helping you.

There are three types of weather briefings: Outlook, Abbreviated, or Standard. Pilots often do not get all the information they need because they do not use the proper terms when asking.

Ask the briefer for an "Outlook" briefing when the time of your proposed takeoff is more than six hours from the time of the briefing. The briefer will then give you basic forecast information. If you are limited to VFR weather, the briefer can give you a good idea of whether the flight is possible. This type of information should be considered as only a first look at the weather for your flight. Later, when the time to fly is at hand, you will need to call again for more specific information.

The specific information for flight planning purposes and real decision making is given when you ask for a "Standard" briefing. This briefing will be an "all-there-is-to-know" discussion of the weather. When the briefer starts talking, sometimes it is hard to keep up with him. It is easier if you know the format. Here is how the briefing is conducted:

1. **Adverse Weather Conditions.** The briefing begins with the bad news. It is done this way so that, if within the first few sentences you realize the weather is not good for the flight and the rest of the briefing is pointless, you can stop the briefer and save everyone valuable time. If you hear, "A line of thunderstorms is moving in your direction" you will probably not need to listen to the rest. You get information about SIGMETs, AIRMETs, and Center Weather Advisories here.

2. **VFR Advisory**. If you originally told the briefer you were proposing a VFR flight, the briefer may at this time make the statement, "VFR flight is not recommended." The briefer makes this statement only after becoming fully familiar with the ceilings, visibilities, and weather advisories along your proposed route of flight. As always, the pilot is solely responsible for the flight. The FSS briefer cannot tell you not to fly, but his advice usually should be heeded.

3. **Synopsis.** The synopsis is a broad-brush description of the current weather. It will include type, location, and movement of large weather patterns like fronts and air masses that may affect the flight you intend to make.

4. **Current Conditions.** The briefer will collect information from all sources reporting what is taking place at that time. This includes current weather observations made at your departure and destination airports, observations made at airports along your route of flight, pilot reports in the area of your flight, and current weather radar information. If you call more than two hours before the flight, this information might be omitted from the briefing since this information will be obsolete at the time of the flight.

5. **En route Forecast.** The briefer will look at all available forecast material for your route of flight. Usually he will give you information in chronological order. The briefer will discuss forecast weather at your departure point at the proposed time of departure. Then, he'll discuss en route forecasts pertinent to your flight.

6. **Destination Forecast.** The briefer will give you an outlook for weather conditions during your descent and landing at your destination at your estimated time of arrival.

7. **Winds Aloft.** The briefer will give the predicted winds at various altitudes above the ground. Remember, the winds at the surface and the winds aloft can be very different both in velocity and direction. You should use the winds aloft when determining en route ground speed and fuel consumption. The winds aloft are given at nine different levels from 3,000 feet MSL to

39,000 feet MSL. When a station's elevation is within 1,500 feet of one of these predetermined levels, the wind is not forecast. This is why some stations will have less than nine forecast wind levels. The briefer may assume that you need only the lower wind levels in the briefing. Make sure you ask for other levels if you think they will be considered in your flight planning.

8. **Notices to Airmen (NOTAMs).** Specifically the NOTAM(L) and NOTAM(D) types. The "L" refers to Local information. The Flight Service Station collects NOTAMs in an area within approximately 400 miles of the station. This information includes runways being closed and construction on or near the airport. The "D" refers to Distant Dissemination beyond a particular FSS's 400 mile area of responsibility. This information can include radio navigation facility closures, service interruptions of instrument approaches or radar, or even entire airports being closed. There is a common problem here, however: The briefer will not include "D" NOTAMs if they have already been printed in the "Notice to Airmen Publication." If you do not have the publication, you will need to ask the briefer for NOTAM(D) specifically. Also, FDC (Flight Data Center) NOTAMs are not normally included in a weather briefing. FDC NOTAMs refer to regulation changes or changes to aeronautical charts and instrument approaches prior to the next scheduled publication date of the chart. The briefer should be able to give you any NOTAM information if you specifically ask for it.

9. **Air Traffic Control Delays.** If the briefer knows of any reason why the flight cannot be handled by air traffic controllers at the time you propose the flight, he will inform you. This basically pertains to IFR traffic, although IFR congestion at a busy airport would certainly affect a controller's ability to handle VFR traffic. The controllers would much rather have you sit and wait at the end of a runway than have you perform holding patterns in the air. If traffic is routinely being "held for release," the briefer should advise you. A release will come when a "slot" opens up either en route or at your destination.

If you are not getting as thorough a weather briefing as this format indicates, then you are not using the right language. The term "Standard Briefing" unlocks the flood gate of information. In addition to what is in the format, you can ask the briefer for specific information about military training routes, MOAs, restricted airspace, density altitude, stability of the atmosphere, air traffic rules, U.S. customs services, and search and rescue procedures.

If you wanted to check one last item about the weather before your departure, you could call the FSS and ask for an "Abbreviated Briefing." Provide the briefer with the necessary background information, and then ask specific questions. This type of briefing should be requested only after you have received weather information from some other source prior to this call. The other source might have been a previously received Standard Briefing. The Abbreviated Briefing should be seen as a supplemental information source, never as the only information source.

THE NEW AUTOMATED FLIGHT SERVICE STATIONS

The demand for weather information and other services provided to pilots is on the rise. The FAA feels that in order to handle the future demand, some changes must be made. The system that pilots learned prior to and during the 1980s is being replaced by a more consolidated and automated system. From the pilot's point of view, local Flight Service Stations will be closing. The FAA has chosen to take a regional rather than a hometown approach. Four or five Flight Service Stations at a time will be shut down and swallowed up by an Automated FSS somewhere in your area. The FAA has determined that there is less need for face-to-face communications with a briefer. Gone will be the one-on-one human touch, lost in the name of progress. This is not to say that you cannot go and talk personally to a weather briefer, but you may have to travel a long way to do it.

One of the biggest objections raised about Automated FSSs is that computer-generated briefings require the pilot to turn his push button telephone into a computer keyboard. We still do not completely trust computers. When we file a flight plan over the phone to a living human being, we hang up and envision that person going right to work for us. Giving the same information to a computer seems cold, like talking to a telephone voice recorder. We think that it may be a long time until a human gets the message. The FAA is hoping these objections are just temporary problems of acceptance and troubleshooting. The FAA has done and is doing several things to convince pilots that this is the way to go. After the year 2000, there will probably be no need for something called a Flight Service Station. We will all have computer access, and weather information will therefore be readily available in our homes, cars, and airplanes. We already have it in our airports.

The positive side to Automation is access to more, better, and faster information. This will put more burden on the pilot to keep up with the flow. The pilot will have to become an ***information manager***. We now depend on an FSS weather briefer to pick out what bits of information might be pertinent to our flight. In the future, we may have no one on whom to rely but ourselves. We will have to become translators of vast amounts of incoming information. The computer will just be the bank; we must figure out from which account to withdraw.

If the Automated FSS will swallow up many smaller observation stations, who will be making the weather observations in the future? The answer is: robots! Well, not a robot with arms and legs, but an automatic machine capable of taking weather information. Currently the Automated Weather Observation Station (AWOS) is in operation at approximately 90 locations. These stations operate without any human assistance or may have part-time people to assist the robot. AWOS is a solid-state machine that measures and reports temperature, dew point, wind direction and velocity, altimeter setting, and precipitation accumulation, density altitude, visibility, cloud coverage, and ceiling.

The Automated FSS will manage more "mass dissemination" outlets. These mass dissemination outlets include TIBS (Telephone Information Briefing Service), PATWAS (Pilots Automatic Telephone Weather Answering Service) TEL-TWEB (Telephone-Transcribed Weather Broadcast) and DUAT (Direct User Access Terminal Computer). In other words, there will be more ways to get weather information without actually talking to a human. The FAA would like a transition period where pilots use more of the mass dissemination outlets and less personal contact. For example, you get detailed weather from a recorded message over the phone and then for quick details you call the FSS and request an Abbreviated Briefing. If you do this, you should tell the briefer, when you finally reach one, which mass dissemination outlet you have used. This will help the briefer fill in the gaps. This, in theory, will free up the briefer to handle more calls.

IN-FLIGHT

Most pilots get weather information by telephone or DUAT, plan their flights, then get in the airplane and never access the weather information system again. They isolate themselves from the wealth of information that is available on their radio if they only knew how to get it. Pilots even will call in flight plans but be unable to activate the plan once airborne because they do not know several ways to get into the system.

Many VFR pilots use their radios when departing an airport, but then the radio goes essentially unused for the remainder of the trip. Pilots then fly across the land listening to downwind reports at distant airports and idle chatter on the unicom frequency, when vital information passes them by on the FSS frequencies. Before takeoff, write down the FSS frequencies that you can receive along your route of flight. Then while in flight, monitor the frequency that will do you the most good. In this way vital decision-making information will come to you. You will not even have to go scrambling for it!

If the weather broadcast in the AIRMETs, SIGMETs, and CWAs are so important, why are they only broadcast at certain intervals, allowing for the possibility that a pilot would miss an alert? This question is being answered by a new system that brings all these alerts under one umbrella: the Hazardous Inflight Weather Advisory Service (HIWAS). HIWAS is a continuous broadcast that summarizes information from all existing AIRMETs, SIGMETs, CWAs, and PIREPs. When a HIWAS alert is issued, the announcement can be heard on all

ARTCC, FSS, and airport terminal frequencies. The announcement will instruct airborne pilots to contact the continuous HIWAS frequency in their area. The pilot can then switch frequencies (or use the "both" feature of the audio panel) and hear the recorded message. HIWAS is not yet a nationwide service. In areas where HIWAS has been installed, the local FSS and ARTCCs will stop broadcasting alert messages at time intervals around the clock and rely on the HIWAS system. In your pre-flight briefing, you should ask if the area along your route of flight has HIWAS service.

If you are monitoring a frequency and an alert is broadcast, your next step is to talk with someone. With the exception of an FSS phone call, all the services discussed thus far involve a pilot listening to a recording or a computer-generated voice. But to get the best information, you will have to ask questions that are unique to your situation. If hazardous weather is in your area or along your route of flight, you will have the classic decision. You must decide whether or not to continue the flight as planned, stop at an interim airport, or turn around and go back to where you came from. Federal Aviation Regulation part 91.5 speaks to the required pilot action in the event that something causes you to consider changing plans: *"For an IFR flight, or for a flight not in the vicinity of an airport... [the pilot must familiarize himself with] alternatives available if the planned flight cannot be completed."* If while in flight you hear a weather alert or actually see bad weather ahead, you need information immediately to help you make your decision.

The best thing to do is to talk to someone who has access to real-time information. If the situation concerns precipitation and/or thunderstorms, you want to talk to someone who is watching a radar screen. Your first attempt at communicating should be to a Flight Service Station. The easiest way to contact a Flight Service Station is on a "discrete" frequency. A discrete frequency means that you dial in the correct numbers and broadcast, then in return the person at the FSS talks back to you on that same frequency. A control tower is another example of a discrete fre-

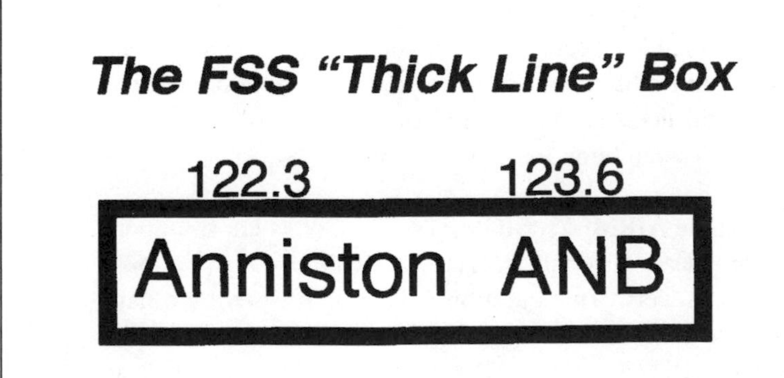

Fig. 14.2

quency. These frequencies can be found on the top on a thick-lined FSS information block (figure 14.2) on a sectional chart. If an FSS and a VOR are co-located, then the entire VOR information block is outlined with a thick line. The frequencies of the FSS are located outside the block on the top thick line. The discrete frequencies are those without the letter **R** beside them. Every Flight Service Station has 122.2 and 121.5 as standard frequencies. So even if you could not find the information block, you could try to contact an FSS on 122.2. The emergency frequency is 121.5, and all FSSs listen in on that channel as well.

When you speak to the FSS with the discrete frequency tuned in, you can address them with the term *Radio* followed by the name of the FSS. The briefer will reply back to you on the same frequency. Now you can get the valuable information on which to base your decision. What if, due to the line-of-sight limitations of the VHF radio, you are unable to contact the briefer on the discrete frequency? The answer is certainly not to give up trying to get a personal briefing. You must now fall back to plan B. The FSS system anticipated situations where you would be out of range and designed ways in which the long arm of the FSS could be extended. The range of the FSS and therefore your ability to receive information in flight is extended at certain VOR stations. When a VOR is capable of providing a communications link from your position to the FSS, the VOR information block on the sectional chart will indicate a "parent" station. Under the VOR block is a bracket. Inside the bracket is the name of the Flight Service Station that monitors that VOR. Figure 14.3 shows the top of the box, and a frequency is shown with the letter **R** beside it. The R stands for receive. In other words, the FSS can receive your transmission if you talk on the frequency indicated. For this example the frequency of 122.1 would be placed in the Communications side of the radio and 115.5 in the Navigation side of the radio. When you do this, your transmission does not have to travel a long distance to the location of the FSS but only to the nearby VOR. Your voice is then relayed

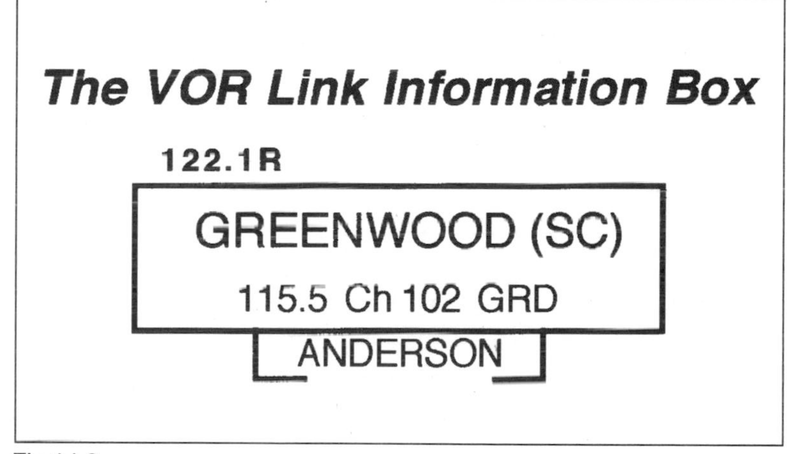

Fig.14.3

by a "land line" to the FSS. The land line is not subjected to line-of-sight limitations and the message gets through. When the briefer responds to your request, he will talk back to you on the navigational frequency of the VOR. Again his voice travels the land line to the VOR and then through the air to your airplane. You must remember two things when using this link: Mention to the briefer on which VOR you would like him to respond to you and turn up the volume on the Navigation side of your radio. Now you talk on one frequency, and the briefer talks back on the VOR. The valuable communication has now been established over a longer distance.

What if the VOR is out of service or there is no VOR in range that provides the link? You must develop another plan. Fall back to plan C. Many Flight Service Stations in addition to monitoring VORs will monitor Remote Communications Outlets (RCOs). The RCO is another radio/land line combination, but it is a discrete frequency and has no navigational function. The RCO is illustrated on sectional charts by a thick blue box with location of the RCO antenna and the letters *RCO*. Figure 14.4 shows an RCO box. The RCO frequency is on the top and the parent station is in the bracket below. When you call an RCO frequency, your voice is transmitted through the air to the RCO antenna. Then your voice takes the land line the rest of the way to the FSS. The briefer returns the favor using the land line for the first part of the trip back to the RCO site and then back through the air to your radio. The actual "through the air" part of the transmission can be up to 50 nautical miles, depending on the line of sight. As the Flight Service Stations consolidate, there will be even more reliance on VOR and RCO communication links.

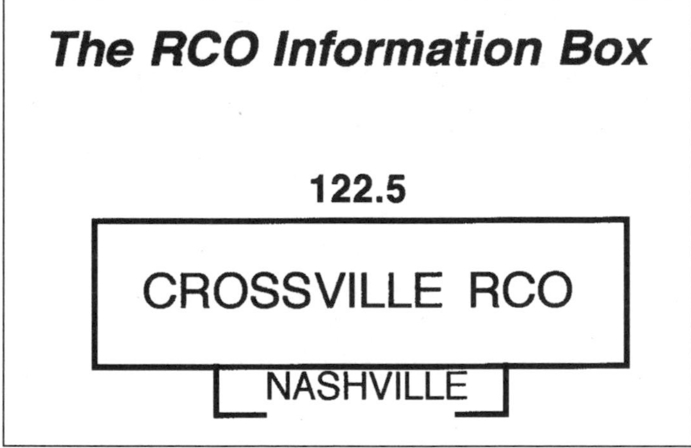

Fig. 14.4

The Air Route Traffic Control Centers use RCOs as well. The Center controller may be controlling traffic in an area that is hundreds of miles from the actual location of the center. Just like the FSS, the controller cannot communicate using line-of-sight transmissions across those miles to the place where his radar screen has coverage, so he uses an RCO that is located somewhere under the radar coverage.

With the direct communication possibilities to a Flight Service Station, together with VOR and RCO relays, the network of coverage is almost nationwide. If you are approximately 3,000 feet AGL or higher, you should be able to use some method to reach weather information. Someday all these communications will be delivered via satellite, and all the present communication links will no longer be necessary. Then, rather than using land lines to carry voices over mountains and around the curvature of the Earth, we will be able to transmit and receive through an orbiting relay station.

If for some reason plans A, B, and C fail to reach a Flight Service Station, then call the en route specialist: The Enroute Flight Advisory Service (EFAS). EFAS goes by another name as well: Flight Watch. The Flight Watch briefer is an expert in enroute weather; therefore, you should not try to open or close a flight plan on this frequency. The Flight Watch information is no longer shown on sectional charts because now the service is standard nationwide.

Any time you are flying over the United States at 5,000 feet AGL or higher, you can reach Flight Watch. For flights below 18,000 feet MSL, Flight Watch has a universal frequency: **122.0**. If you are not sure which Flight Watch station you are closest to, just call on 122.0 and say "Flight Watch" with your aircraft number. Basically the area in which a Flight Watch station has jurisdiction is the same as the ARTCC areas. The Flight Watch specialist is located at an FSS but can communicate by way of RCO over a wide area of responsibility. Flight Watch is usually operated from 6 a.m. until 10 p.m. every day. Complete diagrams of Flight Watch stations and their coverage areas are shown in the Airport/Facility Directory (figure 14.5).

The Flight Watch specialist is the "teller" at the bank of Pilot Reports (PIREPs). PIREPs are firsthand pilot report information because more often than not that specialist is the person who took the pilot report in the first place. The Flight Watch specialist will often solicit a PIREP from you when you call for information.

When you talk to Flight Watch, you are talking to someone who is looking at all the weather information possible. If you see bad weather ahead on your route of flight, call Flight Watch because the person at the other end is already looking at that bad weather on radar. The Flight Watch specialist is in the best position and has the best information to give you decision-making data. How wide is the bad weather pattern? Which way is it moving? Can I go around the weather safely? If I can go around, which direction is best? When you get the answers to these questions, it will be easier to make important weather-related decisions. Will you turn around, stop at an airport nearby and wait out the weather, or continue on? The decision is best made with all the facts.

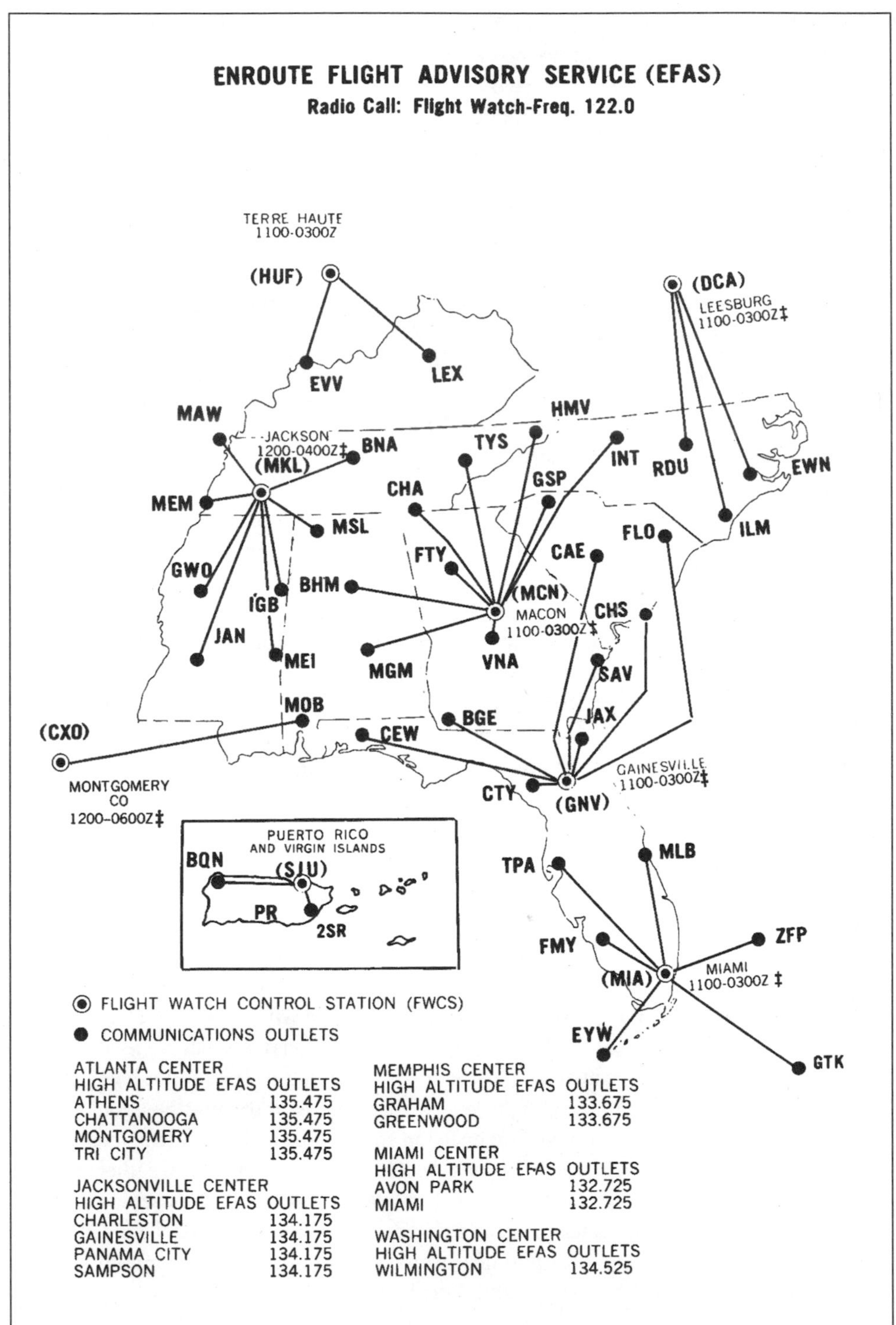
ENROUTE FLIGHT ADVISORY SERVICE (EFAS)
Radio Call: Flight Watch-Freq. 122.0
TERRE HAUTE
1100-0300Z
(HUF)
EVV
LEX
(DCA)
LEESBURG
1100-0300Z‡
MAW
JACKSON
1200-0400Z‡
(MKL)
BNA
HMV
TYS
INT
RDU
EWN
GSP
CHA
MEM
MSL
ILM
FLO
CAE
FTY
GWO
BHM
(MCN)
MACON
1100-0300Z‡
CHS
IGB
JAN
MEI
MGM
VNA
SAV
MOB
BGE
JAX
(CXO)
CEW
MONTGOMERY
CO
1200-0600Z‡
CTY
(GNV)
GAINESVILLE
1100-0300Z‡
PUERTO RICO
AND VIRGIN ISLANDS
BQN
(SJU)
PR
2SR
TPA
MLB
FMY
ZFP
(MIA)
MIAMI
1100-0300Z ‡
EYW
GTK
FLIGHT WATCH CONTROL STATION (FWCS)
COMMUNICATIONS OUTLETS
ATLANTA CENTER
HIGH ALTITUDE EFAS OUTLETS
ATHENS 135.475
CHATTANOOGA 135.475
MONTGOMERY 135.475
TRI CITY 135.475
JACKSONVILLE CENTER
HIGH ALTITUDE EFAS OUTLETS
CHARLESTON 134.175
GAINESVILLE 134.175
PANAMA CITY 134.175
SAMPSON 134.175
MEMPHIS CENTER
HIGH ALTITUDE EFAS OUTLETS
GRAHAM 133.675
GREENWOOD 133.675
MIAMI CENTER
HIGH ALTITUDE EFAS OUTLETS
AVON PARK 132.725
MIAMI 132.725
WASHINGTON CENTER
HIGH ALTITUDE EFAS OUTLETS
WILMINGTON 134.525

Fig.14.5

15
VOR introduction

IN 1946 A SYSTEM WAS INTRODUCED THAT BECAME THE "BACKBONE" of the Radio Navigation system. The system is called the Very High Frequency Omnidirectional Range, or VOR for short. The VOR receiver in an airplane can determine its own position in relation to the VOR station's transmitter. How is this radio so smart? The VOR radio in the airplane actually receives two transmissions at once from the VOR station. The first signal is called a Reference Phase Signal. This signal is broadcast out in all directions from the station. The second signal is the Variable Phase Signal. The variable signal changes its phase in comparison with the reference signal at all directions from the station. The "sine-wave" of both signals match exactly only at a position that is north of the station. At every other location the first and second signals are out of phase, in other words, they do not match.

The VOR radio in the airplane "looks" at the reference signal and compares it with the variable signal (figure 15.1). If the signals match, then the radio knows it is north of the station. If the signals do not match, the radio will then calculate just how far apart the signals are. If the signals are out of match by one-quarter or 90 degrees, the radio knows that the airplane is one-quarter of the way around a circle from the station. Starting at north and moving clockwise around a circle for one-quarter turn, you find yourself at a position east of the station. Every other position around the station can also be determined in the same way. The radio in the airplane can determine where it is in relationship to the VOR station.

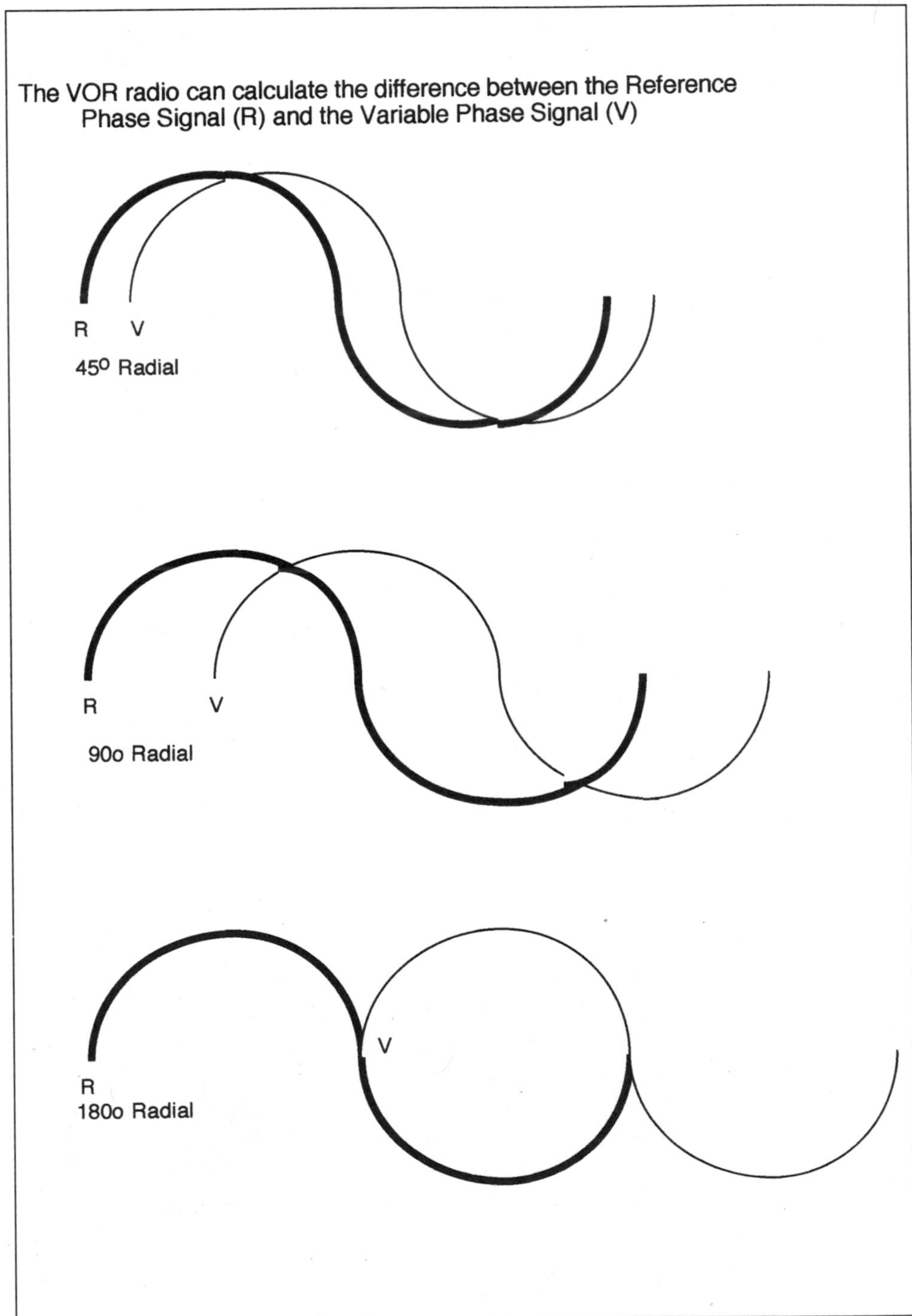
The VOR radio can calculate the difference between the Reference
Phase Signal (R) and the Variable Phase Signal (V)
R
V
45o Radial
R
V
90o Radial
V
R
180o Radial

Fig.15.1

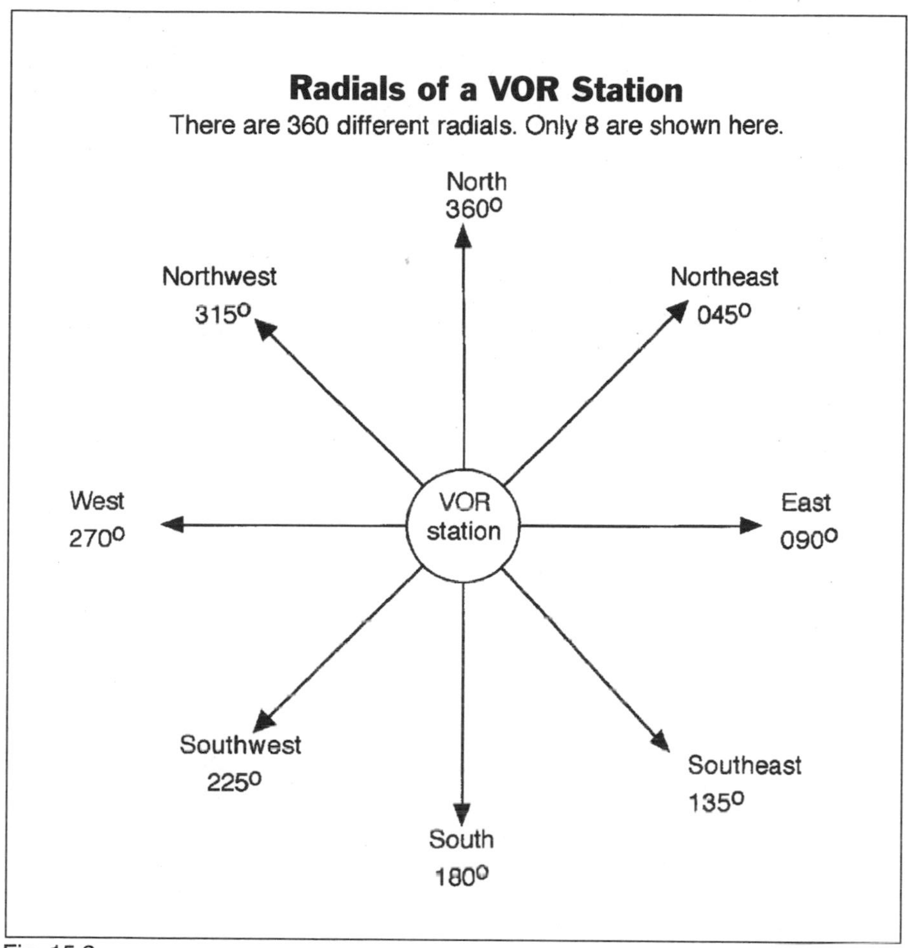

Fig. 15.2

Each position around the circle is called a "Radial." Radials all start at the VOR and "radiate" out in all 360 directions from the station (figure 15.2). Picture the VOR like a bicycle wheel. The spokes of the wheel are the radials. The hub is the VOR station. Each spoke begins at the hub and moves outward. By the time the spokes reach the rim or tire, they have spread out. Radials are the same way. They are packed tightly at the station but spread out with distance from the station. If the VOR radio can tell the pilot which radial the airplane is on, then the pilot will know the airplane's position.

The information is passed on to the pilot by an instrument located on the airplane's instrument panel. But the pilot must learn to interpret the information.

THE VOR EQUIPMENT

The VOR receiver in the airplane consists of two pieces of equipment. First is the radio itself. Many manufacturers build a VOR radio into the same unit as a communications radio. These are called Nav/Com's because they have a VOR navigation side and a communications side in one set. The second component is the VOR "head." The head is a dial that at first glance looks like a Heading Indicator. The degrees of a circle are displayed on the outside of the circular face (figure 15.3). This compass face is called the Omni Bearing Selector (OBS).

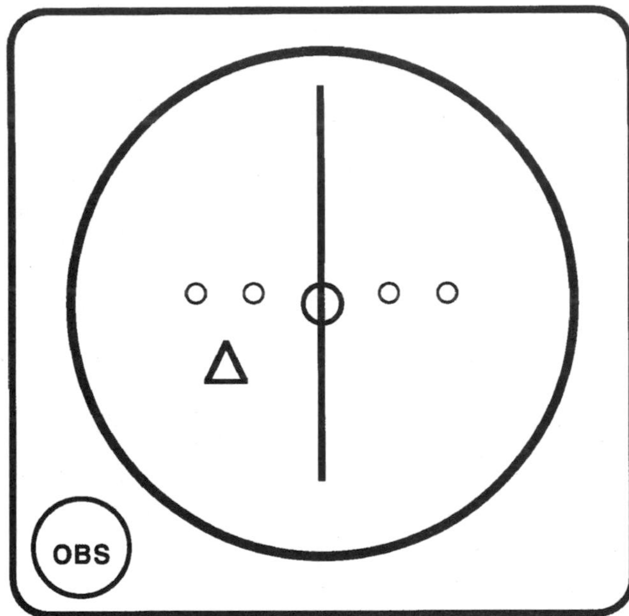

Fig.15.3

Inside the compass is a vertical needle that will swing like a pendulum from the top of the instrument. The needle is the Course Deviation Indicator (CDI). And finally, the face has an arrow that indicates the setting's position "TO" or "FROM" the station.

The pilot can fly directly TO or directly away FROM the VOR station when the OBS is properly set. Refer to figure 15.4. The airplane shown in this figure is flying from the bottom of the diagram to the VOR station. The station's name is Bravo and has a frequency of 115.0. The pilot of this airplane desires to fly inbound to Bravo on the 170° Radial. The 170° Radial begins at the station and moves almost south (the 180° Radial would be exactly south). The pilot wants to fly along the 170° Radial to the VOR, but to do so he or she must fly the opposite direction from the Radial's number. To fly inbound to the VOR on the 170° Radial the airplane must fly a heading of 350°. This is illustrated by the airplane in position number 1 in figure 15.4. The airplane in position 1 is on the 170° Radial, but chooses to fly a heading of 350°. This will lead the airplane to the VOR station. The pilot turns the OBS so that 350° is at the top of the VOR head. ***When the OBS setting and the airplane's heading are the same, then the CDI needle can be "trusted."*** By that I mean that when the CDI needle sways to the right the airplane needs to fly to the right. If the needle moves left the airplane also needs to fly left to remain on course. The airplane at position 1 has a heading of 350° (shown on the airplane's nose) and an OBS setting of 350° (shown below the VOR head of position 1). The needle is centered, which tells the pilot that the airplane is on the course selected.

VOR WIND CORRECTION

In figure 15.4 there is a wind blowing from the East as shown. In time the airplane, while flying a heading of 350°, is drifted to the West of the course line. Position 2 shows the airplane blown off course. The OBS and heading are still the same: 350°, but the needle is no longer in the center of the dial. The needle is off to the right. This indicates that the selected course is to the right of the airplane's current position. At position 2 the pilot turns the airplane to the right from the heading of 350° to a new heading of 020°. The pilot hopes that this new heading will cut into the wind and the airplane will again rejoin the selected course. As the airplane moves from position 2 to position 3 the needle will slowly move from the right back to the center as the airplane intercepts the course.

But this pilot has learned a lesson about the wind. The first time when the heading of 350° was flown, the wind blew the airplane off course. The pilot figures that if 350° is held once again, then the airplane would be blown off again. So instead of flying 350° the pilot tried 360°. This means that the pilot is flying with a 10° wind correction angle!

From position 3 to position 4 the pilot holds a heading of 360°, but as it turns out this 10° wind correction angle is more than required. The CDI needle has moved to the left. The airplane is slightly off course again, but this time the airplane is East of course. Now the pilot looks at the facts. A heading of 350° allowed the airplane to blow off to the left. A heading of 360° caused the airplane to fly off course to the right. What course would "split the difference" and keep the airplane right on course? At position 5 the pilot tries a heading of 355°, which splits the difference between 350° and 360°. The 355° heading is a 5° right wind correction angle. The left heading of 350° and the right heading of 360° forms a "bracket." The pilot now knows that the perfect heading that will keep the airplane on a straight course to the VOR station lies somewhere inside this bracket. The pilot uses 355° because it is in the middle of the bracket.

In all positions from 1 through 5 the OBS has remained set on 350° and this has produced a TO indication. From position 5 to position 6 the airplane passes over the VOR station. When the airplane passes over the station the indication will change from TO to FROM and the airplane will be along a new radial: the 350° radial. The same heading of 355° that was used to hold the course TO the sta-

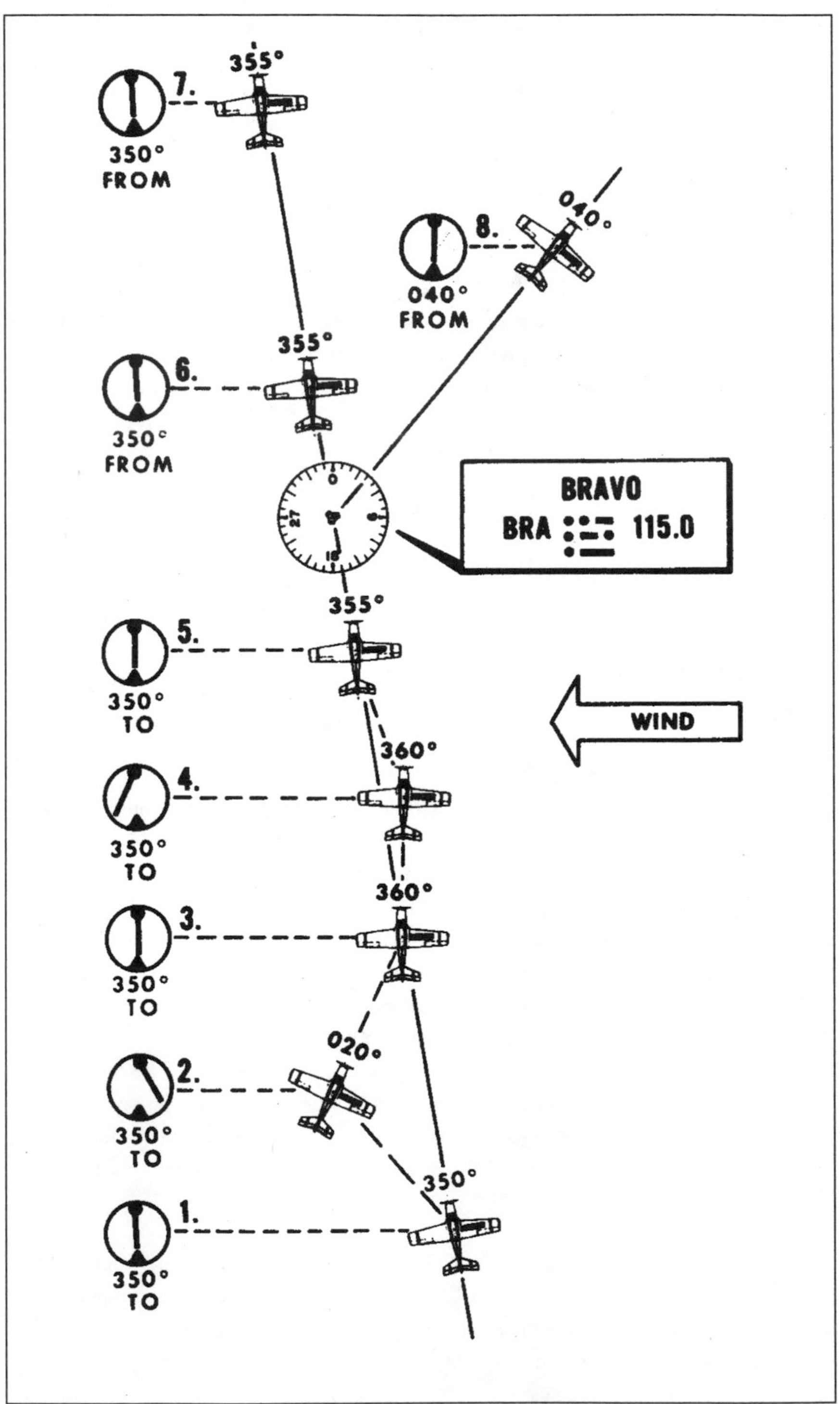
355°
7.
350°
FROM
040°
8.
040°
FROM
355°
6.
350°
FROM
0
9
18
27
BRAVO
BRA 115.0
355°
5.
350°
TO
WIND
360°
4.
350°
TO
360°
3.
350°
TO
020°
2.
350°
TO
350°
1.
350°
TO

Fig.15.4

tion is also holding the course flying away FROM the station from position 6 to position 7.

What if the pilot had wanted to fly off in a new direction after passing VOR station Bravo? Position 8 indicates that the airplane is now flying outbound on the 040° radial. After crossing over the station the pilot turned the OBS knob and placed 040° in the top of the VOR. The pilot then turned the airplane to a heading of 040° and intercepted the course outbound. Even when flying away from the station, the CDI needle can be trusted when the OBS setting and airplane heading are in general agreement. The OBS and heading can be off to allow for wind correction angles and still produce a "fly to the needle" situation.

If the OBS and heading ever disagree, a condition called Reverse Needle Sensing occurs. When the OBS and heading are opposite a "fly away from the needle" indication is produced. The pilot can avoid this distracting situation by simply turning the OBS knob until the OBS setting is the same as the heading.

16
VOR indications and positions

IN CHAPTER 15 WE DISCOVERED HOW THE VOR SYSTEM WORKS, YET actually using the system is still something that pilots have trouble with in the beginning. In this chapter we will get some practice with using the system.

Each VOR station is given its own unique 3-letter identification code. The three letters and the Morse code for those three letters are given in the VOR information box of a chart (see figure 15.4). Before using the VOR the pilot should listen to the 3-letter identifier to be certain that the proper station is being received. The 3-letter identifier in Morse code is played over and over again on the VOR frequency. Since very few pilots actually know Morse code, the pilot should listen to the code while reading the information box. As the dots and dashes are sounded the pilot can verify that the VOR is correct. After verification, the volume can be turned down on the VOR radio. The Morse code identifier does not need to be listened to continuously.

WHERE AM I? USING VOR

After proper identification of a VOR station, how does the pilot use the VOR to determine location? What location would the airplane be in if the VOR had a 180° OBS setting, with a FROM indication and a centered CDI needle? The airplane is South of the VOR station. We know this because the airplane is on the 180° radial. The 180° radial starts at the station and proceeds South of the station. We do not know from the VOR alone just how far from the station that the airplane is, but we do know it is South of the station.

The actual heading of the airplane does not matter. The VOR system only tells location in relation to the station. If the airplane is on the 180° radial and has a heading of 270°, then the airplane will not remain on the 180° radial for very long. Soon the airplane would fly off the 180° radial and over another radial. However, if the pilot chooses to stay on the 180° radial, he or she could turn South and remain on the 180° radial by flying a heading of 180°. Or the pilot could turn inbound and fly North with a heading of 360°.

Here is another example to illustrate that the airplane's heading does not affect the VOR indications. A pilot in flight tunes in a VOR station and identifies the station. The pilot then turns the OBS knob until the needle centers with a FROM indication. With the needle centered, the number 270 is under the VOR index. This means that the airplane is on the 270° Radial or West of the station. The pilot then decides to fly in a circle. While the airplane turns through every heading of the compass during the circle, what will the CDI needle of the VOR do? It will not move! (The needle might move if this took place so close to the station that the airplane's circle was wider than the Radial.) The airplane passes through every heading, but that does not change the fact that the airplane is still West of the VOR station.

How does the TO/FROM indicator work? How can this indicator determine if the airplane is getting closer or farther away from the station, especially if the heading does not affect the VOR indications? Look at figure 16.1a to illustrate this idea. The arrow in this figure represents the OBS setting. If we assume that the top of the page is Magnetic North, then in figure 16.1a, the OBS is set to 360°. Whenever the OBS is set, this divides the Earth into two parts: a TO Zone and a FROM Zone. These two zones are divided by an OFF Zone. What TO/FROM indication would the VOR display if the airplane were in position A or B of figure 16.1a? Recall that in Chapter 15 we learned that "The only time the indications can be trusted is when the OBS setting and the airplane's heading are the same." Therefore, to get any usable information from the VOR we must assume that the airplane at position A or B has been turned and is flying with the same heading as the pictured OBS setting of 360°. If an airplane in position A or B continued on a heading of 360° would that airplane be flying closer TO or farther away FROM the OFF Zone line? It should be clear that flying North from position A or B will take the airplane farther away FROM the OFF Zone line. The airplane does not have to be flying directly away from the station itself, just the dividing line. This VOR would have a FROM indication.

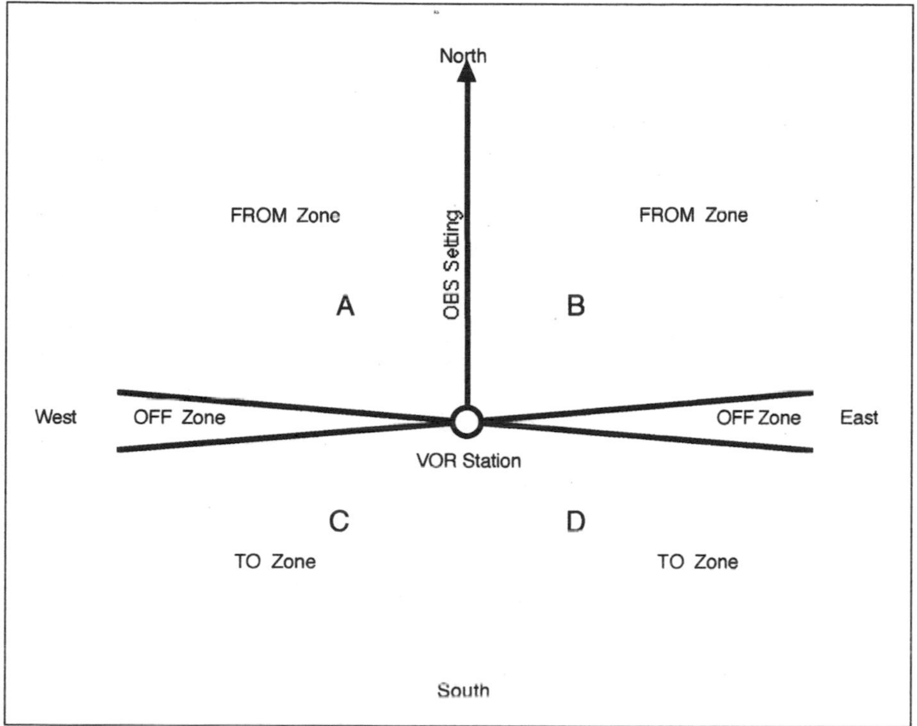

Fig.16.1a

But as you can see, an airplane at position A or B is not on the 360° Radial. Position A is off to the West of the 360° Radial and position B is off to the East of the radial. When the airplane is flying the same direction and the OBS setting is the same, the CDI needle can be trusted as well. It is clear that an airplane flying North from position A must make a right turn in order to fly over to the arrow. Therefore, the CDI needle would be deflected to the right, because the arrow is to the right of the airplane's current position. An airplane at position B would need to make a left turn to intercept the 360° radial and therefore its CDI needle would be deflected to the left.

What would be the indications of airplanes at positions C and D of figure 16.1a? If an airplane at C or D turned and flew a heading to match the OBS setting (360°), would the airplanes be getting closer TO or farther away FROM the dividing line? In each case, C and D, an airplane flying with a heading of 360° would be getting closer TO the line. Therefore, both would have TO indications. Position C would have a right deflected CDI needle because that airplane needs to turn right in order to fly over to the 360° radial. The airplane at position D would have a left deflected needle.

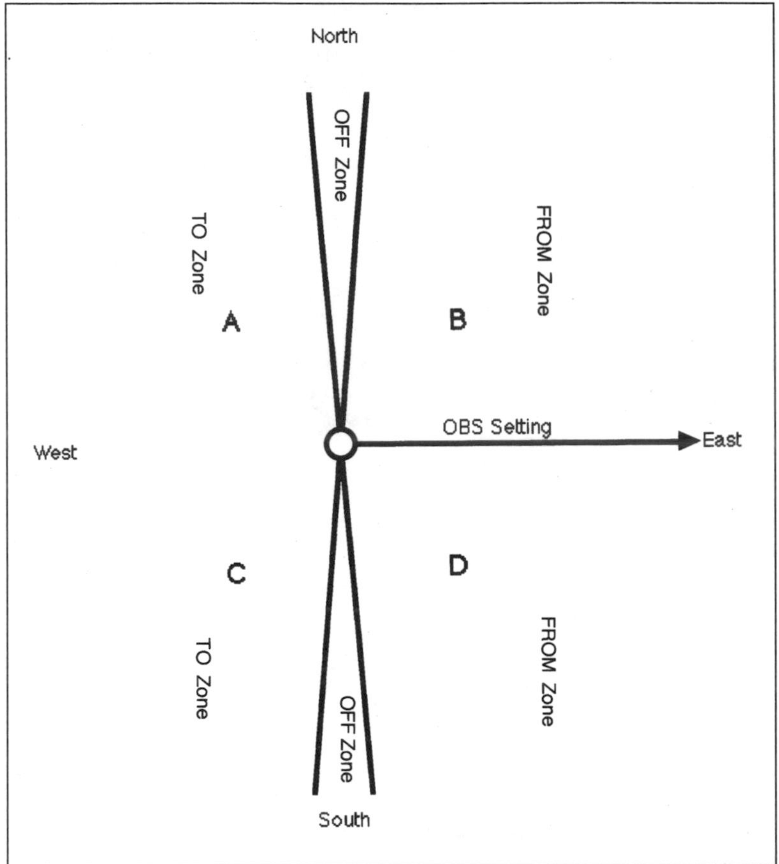

Fig.16.1b

Things change when the OBS setting changes. In figure 16.1b the OBS knob has been turned to 090°. This is indicated by the arrow pointing to the East. Positions A, B, C, and D have not changed, only the OBS was moved. What TO/FROM indication would an airplane at position A have with a 090° OBS setting (figure 16.1b)? Remember, trust the indications only when the heading and OBS setting match, so imagine the airplane at position A flying a heading of 090° now. An airplane flying from position A on a heading of 090° would be traveling closer TO the dividing line. The indication would be TO with a right deflected needle.

Position A did not change from figure 16.1a to figure 16.1b. Only the OBS setting changed between the two figures. So it is possible for an airplane in the same location to have both a FROM indication or a TO indication. It all depends on the OBS setting. What will be the indications at B, C, and D in figure 16.1b?

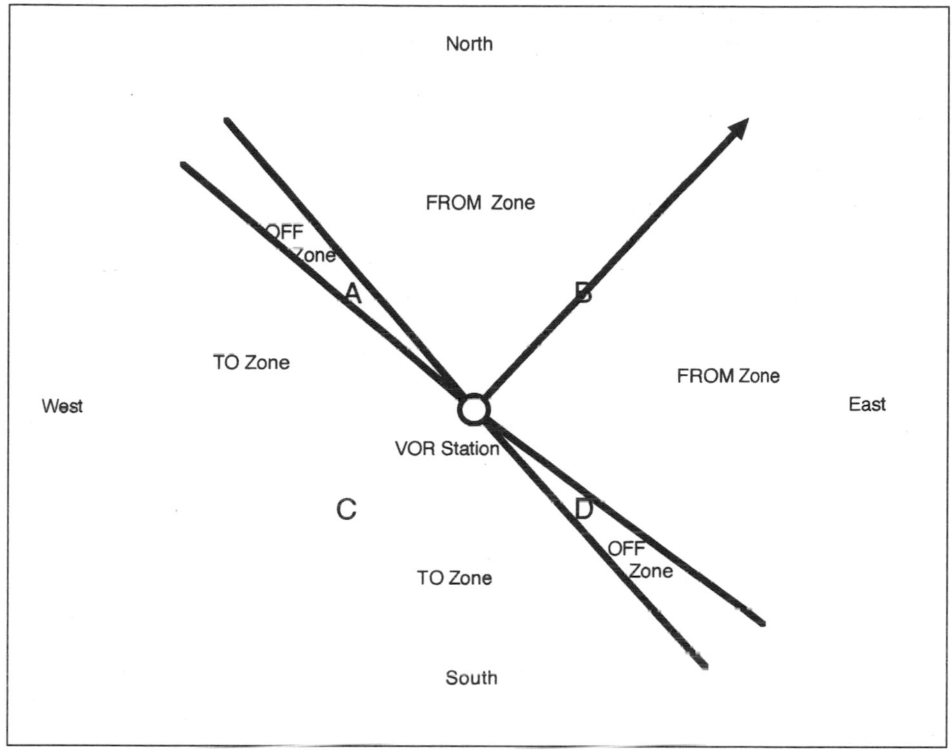

Fig.16.1c

Figure 16.1c presents another situation. The OBS setting in this figure is at 045° as indicated by the arrow. This positioning of the OBS moves the OFF Zone over position A and D. This situation can lead to some misunderstandings.

The TO/FROM indication can also read OFF. Unfortunately the OFF indication can mean two different things. First, if the OFF indication appears it can mean that the VOR station is not being received. Second, it can mean that the VOR station is being received, but the airplane's position falls within the OFF Zone. The best clue for the pilot is to realize that the CDI needle will fall to the center when the VOR signal is not received. So if the needle is centered with an OFF indication a lack of signal should be suspected. However, if the CDI needle is deflected either left or right while there is an OFF indication, this means that the signal is received but the airplane is "***Abeam*** the OBS setting." Abeam means that the airplane is on a 90° angle to the OBS arrow.

In figure 16.1c, an airplane at position A would indicate OFF with a right deflected needle. The indication would not remain OFF for very long if the airplane were flying 045°. It would soon change to a FROM indication. But while in position A the airplane is abeam the OBS setting and indicating OFF while on the dividing line. If the airplane at position A were flying a heading of 045°, the pilot would need to make a right turn to intercept the 045° Radial arrow. Position D would read OFF with a left deflected arrow.

Positions B and C of figure 16.1c need no turns to be along the OBS arrow. Both would have a centered needle. But position B would read FROM, because flying a heading of 045° from position B would take the airplane farther away

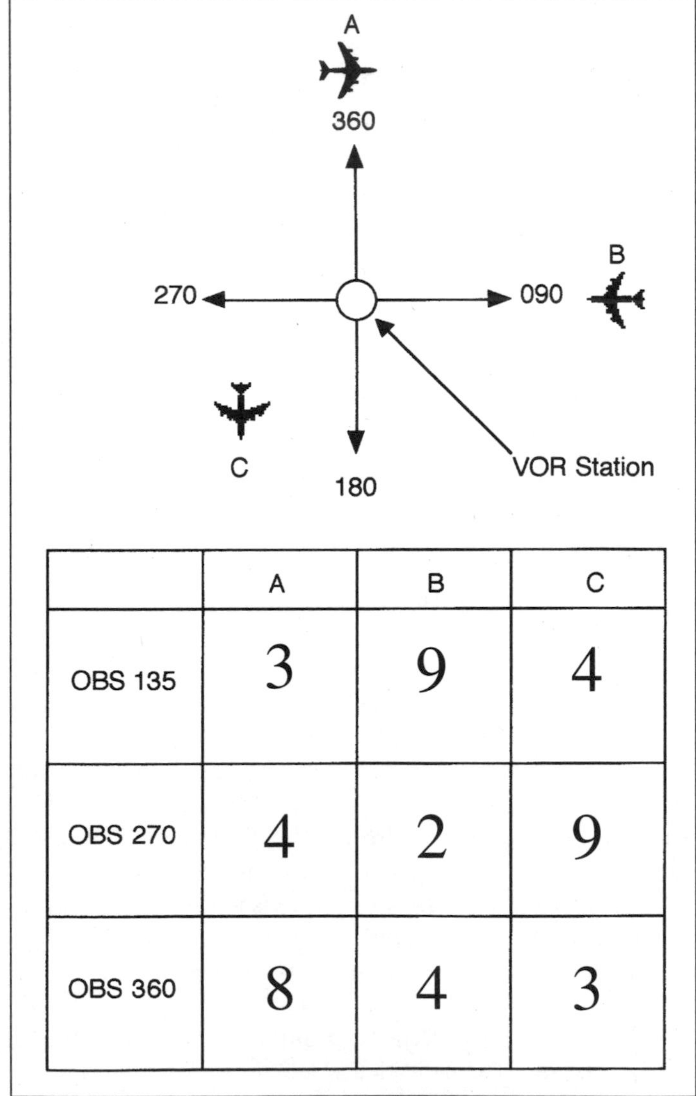

	A	B	C
OBS 135	3	9	4
OBS 270	4	2	9
OBS 360	8	4	3

Fig.16.2a

FROM the station. Position C would have a TO indication because a heading of 045° at position C would lead the airplane closer TO the station.

Now put this knowledge to work. Figure 16.2a displays three different airplanes flying within the signal range of a VOR Station. Airplane A is on the 360° Radial, airplane B is on the 090° Radial, and airplane C is on the 225° Radial. The first column of the table below the diagram pertains to airplane A. If a pilot were flying in position A and turned the OBS setting to 135°, what would be the VOR indication?

Figure 16.2b has a selection of 9 possible combinations of CDI needle positions and TO/FROM/OFF indications. Which of the 9 VORs would the pilot see when in airplane A with a 135° OBS setting? Place the number of the correct VOR in the box under A and beside OBS 135.

Then the pilot, while still in airplane A, turns the OBS to a setting of 270°. Which one of the 9 possible VORs will the pilot see now?

Verify the entire table for all airplanes and all OBS settings!

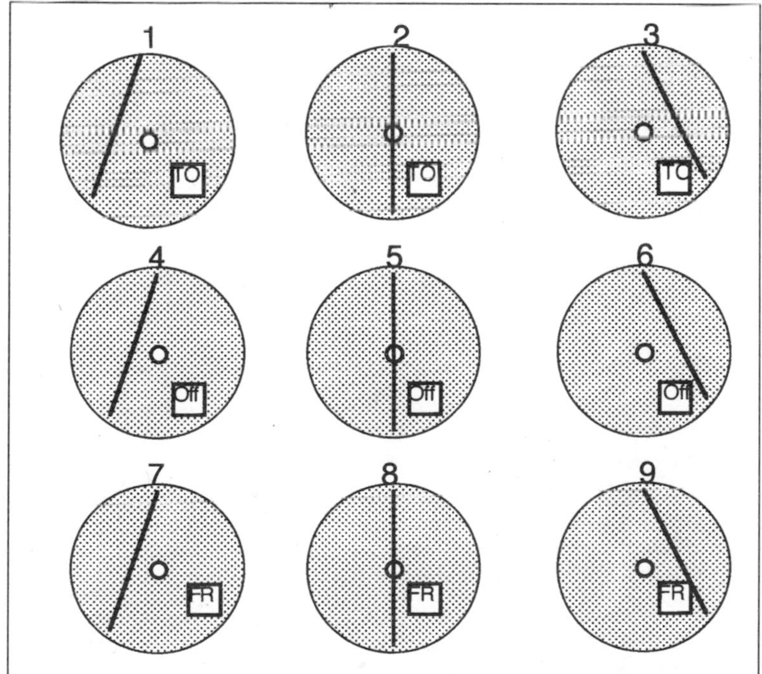

Fig.16.2b

17
VOR receiver checks and DME

IF THE PILOT INTENDS TO USE THE VOR SYSTEM WITH ACCURACY, THE system's ranges and checks must be understood. There are actually three types of VORs: Terminal, Low Altitude, and High Altitude.

The Terminal VOR is a low power station that serves as the navigation aid for an instrument approach to a single airport "terminal." These stations have a range of only 25 miles out and 12,000 feet up. For this reason Terminal VORs cannot be used for enroute navigation and are not used to make up airways. All VORs except Terminal VORs are depicted on the Sectional Chart with a compass rose around the station. Terminal VORs do not have the compass and are indicated on the charts with only the VOR chart symbol and the VOR information box.

The Low Altitude VOR Stations are stronger than Terminals but still only insure an interference-free signal for 40 miles and up to 18,000 feet. The High Altitude VORs have the greatest signal strength. These stations make up the airway system and can be received between 130 and 200 miles in distance. They can also be used as high as 60,000 feet. All three VOR types provide the same in-airplane indications to the pilot.

The VOR signal is broadcast in the VHF frequency band. One of the characteristics of this band is that signals must be "line of sight." This means that there can be nothing standing between the VOR transmitter and the airplane's antenna to block the signal. If a mountain ridge stands between the VOR station and the airplane, the signal will not get through and the VOR receiver will read OFF with no Morse code identification. The VOR station must have a wide open area to operate within in order to work properly. This is why many VOR stations are placed on mountain tops and named for the mountain. Hinch Mountain VOR, Barretts Mountain VOR and Sugarloaf Mountain VOR are examples.

In some locations where VORs are needed, there simply is no clear area. The signal from a conventional VOR can be blocked by tall buildings or terrain, leaving areas where VOR Radials are unusable. To overcome this problem the Doppler VOR was developed. This system reduces the blocking problem in congested areas and needs much less space to operate within. Again the in-airplane indications are the same with Doppler VOR and with all others.

NORTH CAROLINA

VOR RECEIVER CHECK POINTS

Facility Name (Arpt Name)	Freq/Ident	Type Check Pt. Gnd. AB/ALT	Azimuth from Fac. Mag	Dist. from Fac. N.M.	Check Point Description
Fayetteville Muni/Grannis Fld	108.8/FAY	G	005	1.0	On runup area rwy 22.
	108.8/FAY	G	275	0.6	On runup area rwy 04.
Hickory Muni	110.8/BZM	A/2200	225	10.2	Over apch end rwy 24.
Greensboro (Piedmont Triad Intl)	116.2/GSO	G	036	3.5	On twy F.
Kinston (Eastern Regional Jetport at Stallings)	109.6/ISO	G	232	3.3	On end of taxiway to rwy 22.
Lexington Muni	116.2/GSO	A/2300	228	22	Over rotg beacon atop west end of bldg
New Bern (Oak Grove Holf Heliport)	113.6/EWN	A/1100	262	10.6	Over intersection of N-S and E-W rwys.
Raleigh-Durham	117.2/RDU	G	244	1	At end of taxiway to rwy 05.
Simmons	109.8 /FBG	A/150	043°	5.5	Over fire twr.
Sugarloaf Mountain (Asheville Regional)	112.2/SUG	A/2500	280	13.6	Over atct.
Tar River	117.8/TYI	A/1500	260	5.8	Over smoke stack at power house.
Winston Salem (Smith Reynolds)	116.2/GSO	A/2000	297	13.5	Over air traffic control tower.

VOR TEST FACILITIES (VOT)

Facility Name (Arpt Name)	Freq.	Type VOT Facility	Remarks
Charlotte (Charlotte/Douglas Intl)	112.0	G	

Fig.17.1

TESTING THE VOR FOR ACCURACY

Like any equipment the VOR receivers in the airplane need maintenance from time to time. The accuracy of the indications can "creep" to a point where erroneous readings may be presented. The accuracy of the VOR system can be tested and this test is required for Instrument Flight. The test is not required for VFR flight. However, VFR pilots should conduct the test so that they will know what errors are in their VOR systems and then can allow for these errors. There are five different tests that can be conducted. Three types of these tests are listed in the back of the Airport Facilities Directory (AFD). For IFR flight these tests must be performed and recorded every 30 days. The five tests are:

1. VOT (found in the AFD)
2. Designated Ground Checkpoint (found in the AFD)
3. Designated Airborne Checkpoint (found in the AFD)
4. "Make your own" Checkpoint
5. "One VOR against the other"

The first test is the best and most accurate way to determine the accuracy of the VOR. This test comes from a VOR Test Signal called a VOT. Figure 17.1 is a portion of the VOR receiver check information for North Carolina that can be found in the AFD. According to figure 17.1, North Carolina has only one VOT and it is located at Charlotte/Douglas International Airport. The frequency to use is 112.0. The VOT is not a navigational radio, it is for test purposes only. The signal strength is weak and can only be used while on the airport at Charlotte. There are only two possible indications from a VOT. When the OBS is set to 180° the CDI needle should be centered with a TO indication. When the OBS is set to 360° the CDI should again be centered with a FROM indication. It does not matter where on the airport the test is conducted, all that matters is that the signal is received. If 180° is set in the VOR index and the needle is not centered, the pilot should slowly turn the OBS knob until the needle becomes centered. The pilot then reads what OBS setting is under the index. The allowance for IFR flight is plus or minus 4°. If the needle centered with a 185° setting that VOR would be out of tolerance for IFR flight. The VOR could be used in VFR flight but the pilot should make a 5° compensation. Is the VOR within IFR tolerance when the VOT reads 003° with a centered needle and FROM indication? Yes. The 003° setting is within 4° of 360° and FROM is the proper indication for 360°.

All the rest of the receiver checks use normal navigational VOR stations. Look at figure 17.1 and find the entry for Kinston. The type of receiver check available at Kinston is listed with a "G." The G stands for a Designated Ground Checkpoint. This means that the FAA has established a particular location on the ground at this airport to check the VOR. When using an actual VOR for the test rather than a VOT, the location where the test is conducted is important. The listing describes the actual location where the test is to be done as "On the end of taxiway to runway 22." Painted on the taxiway will be a circle that designates the correct spot. The pilot must taxi the airplane onto that circle. When in that exact location the pilot tunes in the nearest VOR. The listing shows this to be 109.6 and that the 232° Radial passes from the station over the painted circle. If the VOR system is accurate, the CDI needle should be centered when the airplane is sitting in the circle while the OBS is set to 232°. The tolerance for this test is also plus or minus 4° for IFR flight.

The listing in figure 17.1 for Winston Salem has the code "A/2000." This means that the VOR receiver check available at Winston Salem is conducted in flight at 2,000 feet. The exact location is "Over Air Traffic Control Tower." To perform this test the pilot must climb to at least 2,000 feet and fly directly over the Control Tower. As the pilot passes over the Tower the OBS must be set to 297°. If the VOR is accurate the needle will center just as the airplane crosses over the Tower. If the needle is not centered the pilot must quickly turn the OBS until it is centered and read the new OBS setting. The setting must be within plus or minus 6° of 297° in this case. There is a greater tolerance for error here to allow for the fact that the airplane is in the air and it is harder to determine exact position.

As you can see from the AFD in figure 17.1, there are only 12 designated VOR receiver checks in the entire state of North Carolina. What if you do not happen to be at one of these designated locations and need to test your VORs? In that case you can use a test that I call "Make your own." To make your own VOR receiver check, take a Sectional Chart and find your nearest VOR station. Then locate a prominent landmark that is approximately 20 miles from the VOR station. Draw a line from the VOR to the landmark and note on which VOR Radial the landmark is located. Now go fly over the landmark. Before crossing the landmark set the OBS to the proper Radial. The needle should center as you pass over the landmark. The tolerance once again for this airborne check is plus or minus 6°.

Last, the final possible check is the "One VOR against the other" test. To perform this test the airplane must have two VOR receivers that are independent of each other except for their antennas. Tune in both VORs to the same VOR station. Center both CDI needles. The OBS settings should now be the same. If they are not the same, they must be within 4° of each other to pass an IFR test. As you can see, this is the least accurate test. The VORs may be within 4° of each other but each could have a 10° error! Use this test as a last resort.

For IFR flight the results of these tests must be recorded on some form available to the pilot in the airplane. When a VOR receiver check is recorded, the person performing the check must write down their name, the date, the place and type of test, and the errors of the VORs. Again, these tests and records are not required for VFR flight, but the aware VFR pilot will test the VORs anyway to determine just how good the VOR information will be.

Type of Receiver Check	IFR Tolerance
VOT	+ or - 4°
Designated Ground	+ or - 4°
Designated Airborne	+ or - 6°
Make your own	+ or - 6°
One VOR against the other	+ or - 4°

When using the VOR alone, the position in relation to the system can be determined but the distance from the station is not known. A pilot may know that he or she is on the 060° Radial, but not know if they are 2 miles from the station or 20 miles out. To solve this problem a system was developed to be a companion to the VOR. This system is called Distance Measuring Equipment, or simply DME.

DISTANCE MEASURING EQUIPMENT

The DME system can determine the distance between the airplane and the VOR station. The DME on the airplane send out a signal which is received by the VOR station. The station in turn relays a signal back to the airplane. The DME in the airplane uses the elapsed time interval between the time the signal leaves the airplane until a signal is received to determine the distance the signal traveled.

Remember back in Chapter 8 we learned Time-Speed-Distance calculations. If any two variables were known, the third variable could be computed. The DME uses the same principle. The DME already knows the speed in which the signals travel. That speed is approximately 186,000 *miles* per second, which is the speed of light. The time portion of the problem is determined by the elapsed time interval or the time it takes the signal to leave and return. With Speed and Time known, the DME solves for Distance. This distance is then displayed for the pilot to see on the instrument.

Some DME systems monitor the location of the airplane from one moment to the next and compare the change in location versus time. This allows the system to calculate the airplane's Groundspeed and predict the Flight Time to the station.

By using the VOR and DME in combination the pilot can pinpoint a location. If the VOR determines that the airplane is on the 060° Radial and the DME determines that the airplane is 12.7 nautical miles from the station, the pilot now knows exactly where the airplane is located.

The only problem that exists with DME is that the distance presented is given in "slant range" nautical miles. Since the distance is determined through signals in the air rather than across the ground, the mileage is "air miles." This is usually not a problem except when flying very near the station. If an airplane flies over the station at an altitude that is 6,000 feet higher than the station, the DME will never read closer than 1.0 nautical miles. The airplane will be over the station, but the DME would never read 0.0 nautical miles unless the airplane crashes into the VOR station!

Practice VOR Receiver Tests (figure 17.2).

Reference the record for each of the following VOR Receiver checks. Use Figure 17.1 to get information on these checks.

1. You are on the ground in Greensboro, North Carolina. You are taxiing on taxiway F. You see the following VOR indications. Is the VOR within IFR tolerance?

Name: P CRAIG

Date: 01-01-97

Place: Greensboro Ground Check

VOR Error: −4°

Within IFR Tolerance? (YES) NO

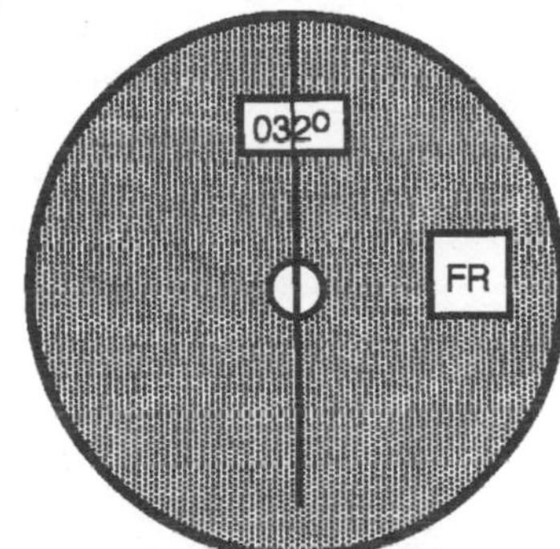

2. You are flying over the Fire Tower near the Simmons Airport. Is this VOR within IFR Tolerance?

Name: P CRAIG

Date: 02-01-97

Place: Simmons Airborne Check

VOR Error: +7°

Within IFR Tolerance? YES (NO)

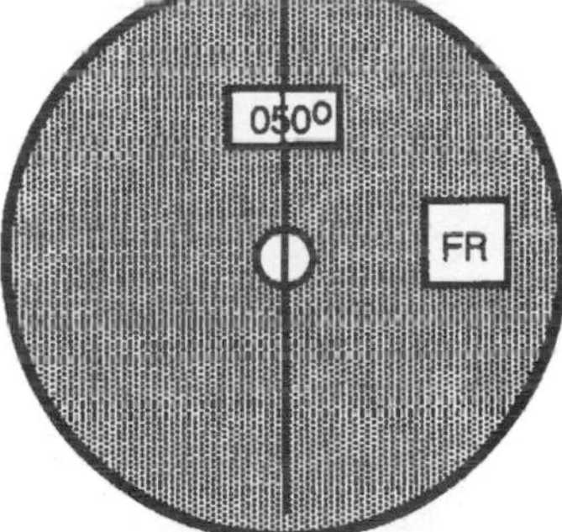

3. You are at Charlotte/Douglas International airport. Your VOR is tuned into 112.0. Is this VOR within IFR tolerance?

Name: P CRAIG

Date: 03-01-97

Place: Charlotte VOT

VOR Error: −3°

Within IFR Tolerance? (YES) NO

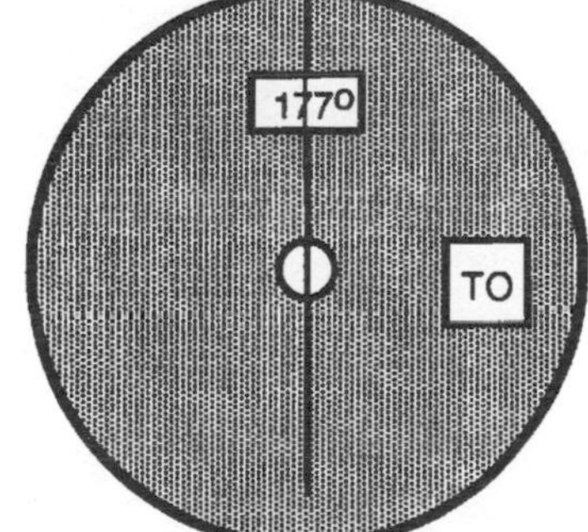

Fig.17.2

18
VOR and lost procedures

EVERY PILOT HAS BEEN LOST AT SOME TIME IN THEIR FLYING CAREER. Getting lost in an airplane is not a comfortable feeling. You cannot simply pull over and ask directions or spread the map out on the cowling. Since becoming temporarily "misplaced" is inevitable, every pilot should also have the skills to get un-lost.

There are several things that a pilot can do to reduce the chances of getting lost. The first line of defense is the Navigation Record. If the pilot plans well and keeps track of the airplane's progress all through the flight the pilot may never get lost. Using checkpoints and timing the legs alerts the aware pilot to the fact that he is getting off course or the groundspeed is different than predicted. Flying without careful preparation can get any pilot lost. If the pilot does get lost, there are two dangers. First, when the pilot does not know the airplane's location it is possible for the airplane to wander into airspace where it should not be. If a pilot got lost and inadvertently crossed into a Prohibited Area, a Restricted Area, or Class B, C, or D airspace without communications, the pilot is placing the flight into danger. Second, while the pilot works to regain position knowledge, the airplane may not be flown with full attention. This could lead to an unexpected stall and poor collision avoidance scan. Last, the airplane will eventually run out of fuel. Making an off-airport landing has many dangers.

IF YOU GET LOST — USE VOR TO GET FOUND

If, after properly planning the flight, the pilot still gets lost, the next best way to get found is to use the VOR. The VOR receiver in the airplane has the ability to determine a position in relation to the VOR station. One of the best ways to get oriented after becoming lost is to use the VOR to locate the airplane's position.

To find a location with VOR only, the pilot must be in a position to receive at least two VOR stations. This may mean that the pilot should climb to a higher altitude in order to get reception from more than a single station.

Figure 18.1 illustrates the VOR cross check. The airplane is high enough to receive two VOR stations. The number 1 VOR receiver is tuned to the frequency 109.6. The number 2 VOR is tuned to 114.1. Both VORs are properly identified by the pilot. The pilot turns the number 1 VOR's OBS knob until the needle is centered with a FROM indication. The pilot then reads the OBS setting to be 090°. This means that the airplane is on this VOR's 090° Radial or East of the station.

But a single VOR alone only gives information about direction. The airplane is still lost because the pilot does not know where along the 090° Radial the airplane is located. The airplane could be just 5 miles East of this VOR or 50 miles. The pilot must narrow this down.

Using a second VOR station, the pilot determines the Radial that the airplane is on from VOR number 2. The frequency of 114.1 is tuned in. The OBS knob is turned until the needle is centered with a FROM indication. The pilot sees that the OBS setting is now 360°. The airplane is on the 360° Radial or North of VOR number 2.

The airplane is on the 090° Radial of VOR number 1 and at the same time on the 360° Radial of VOR number 2. The pilot draws a line across the chart from VOR station number 1 through the 090° position on the compass rose and on out across the chart. Likewise, the pilot draws a line from VOR number 2 through the 360° position and on across the chart. Where the two lines cross is the location of the airplane.

The pilot must divide attention between flying the airplane and performing the cross check. The pilot must have also been prepared with a pencil and a large hard surface to draw the lines on. A clipboard does the job nicely.

The VOR cross check is easiest when the airplane has two separate VOR receivers, but many light trainer airplanes only have one. This means that the pilot must determine the Radial from the first VOR and then quickly change fre-

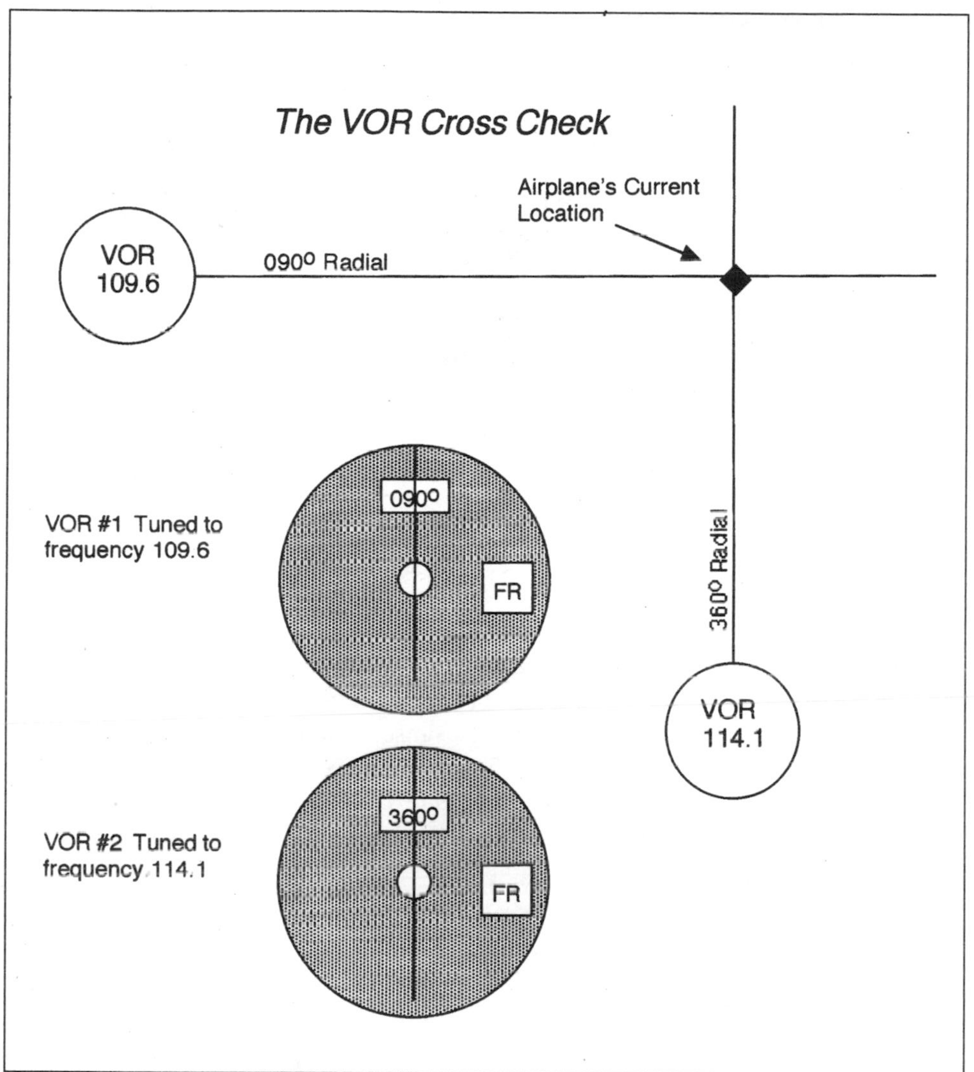

Fig.18.1

quencies and adjust the OBS to determine the Radial of the second VOR. Remember that the airplane is in motion. It does not stay over the same Radial for very long. If there is a long delay between determining Radials the cross check will not be accurate. This is why it is important for pilots to be familiar with this procedure. When a pilot who is not sure how to perform the cross check attempts to do one while in flight, errors occur. This will frustrate the pilot and add to the anxiety of the lost situation. Pilots must know the VOR system well and practice cross checks.

AIR TRAFFIC CONTROL CAN HELP

If a VOR cross check does not solve the unknown position problem and get the pilot back on course, it would be time to confess the situation. Make sure that you include the radio frequencies of RADAR facilities in the areas you will be flying as a part of your normal planning. Write down these frequencies on your Navigation Record. If you get lost and cannot get found on your own call one of these frequencies and ask for help. The controller will ask you to place a four digit code in your transponder. The controller will look across the RADAR scope and will see that transponder code appear. When this happens you will be found! The controller will offer you headings to fly to get back on course. If the transponder code does not appear on that particular RADAR screen, the controller will guide you to call other controllers until you are located.

When lost, remember the three "C's" — Climb, Confess, Communicate.

Never be afraid to call and admit that you are lost. The controller will be glad to assist you. Some pilots have waited to call a controller until they were hopelessly confused, low on fuel, and near sunset. Do not wait to call. Use your Navigation Record first and this should help prevent getting lost. If you get lost anyway, use the VOR cross check (figure 18.2). If that effort fails, call a controller. When lost, remember the three "C's"— Climb, Confess, Communicate.

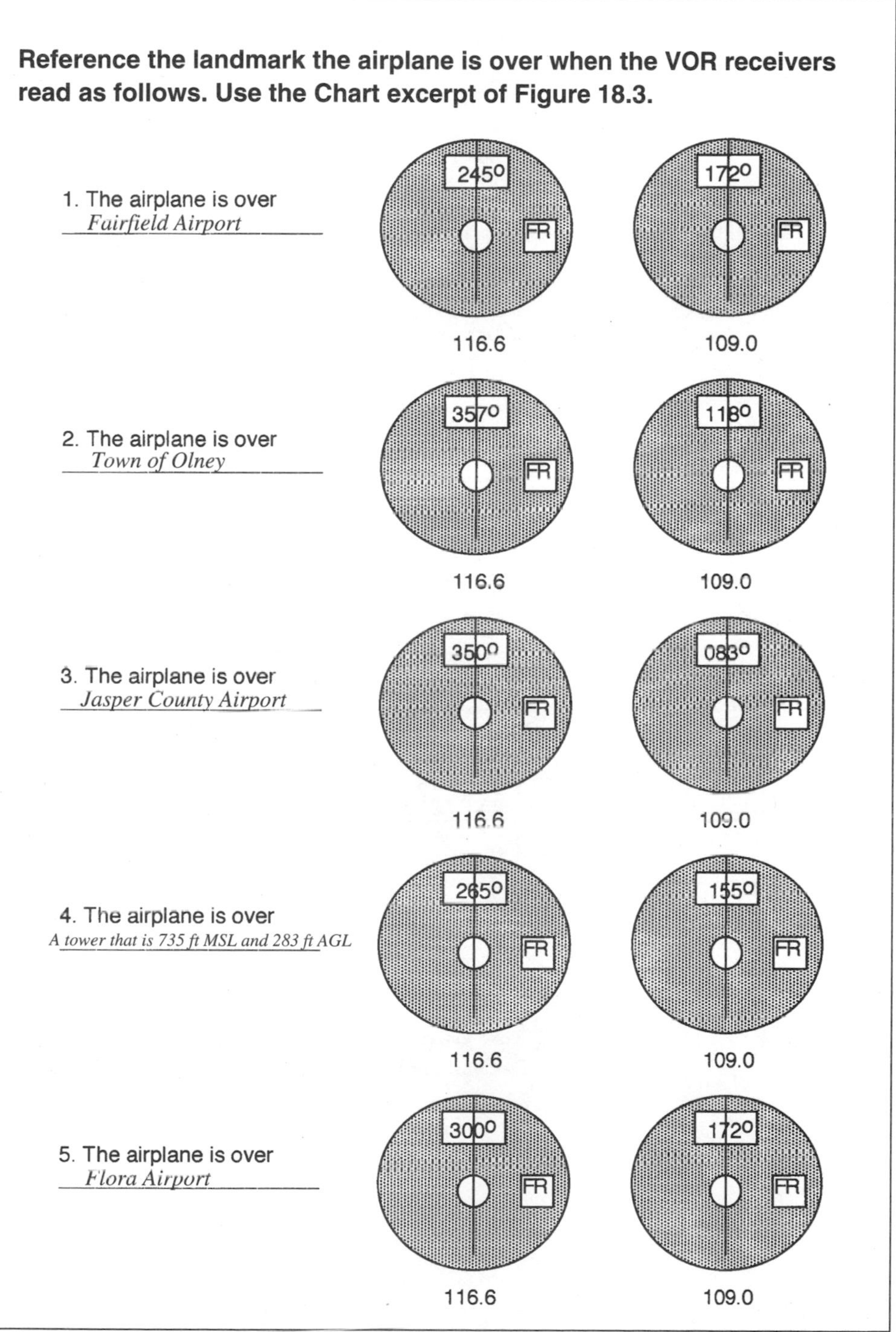
Reference the landmark the airplane is over when the VOR receivers read as follows. Use the Chart excerpt of Figure 18.3.
1. The airplane is over
Fairfield Airport
245°
FR
116.6
172°
FR
109.0
2. The airplane is over
Town of Olney
357°
FR
116.6
118°
FR
109.0
3. The airplane is over
Jasper County Airport
350°
FR
116.6
083°
FR
109.0
4. The airplane is over
A tower that is 735 ft MSL and 283 ft AGL
265°
FR
116.6
155°
FR
109.0
5. The airplane is over
Flora Airport
300°
FR
116.6
172°
FR
109.0

Fig.18.2

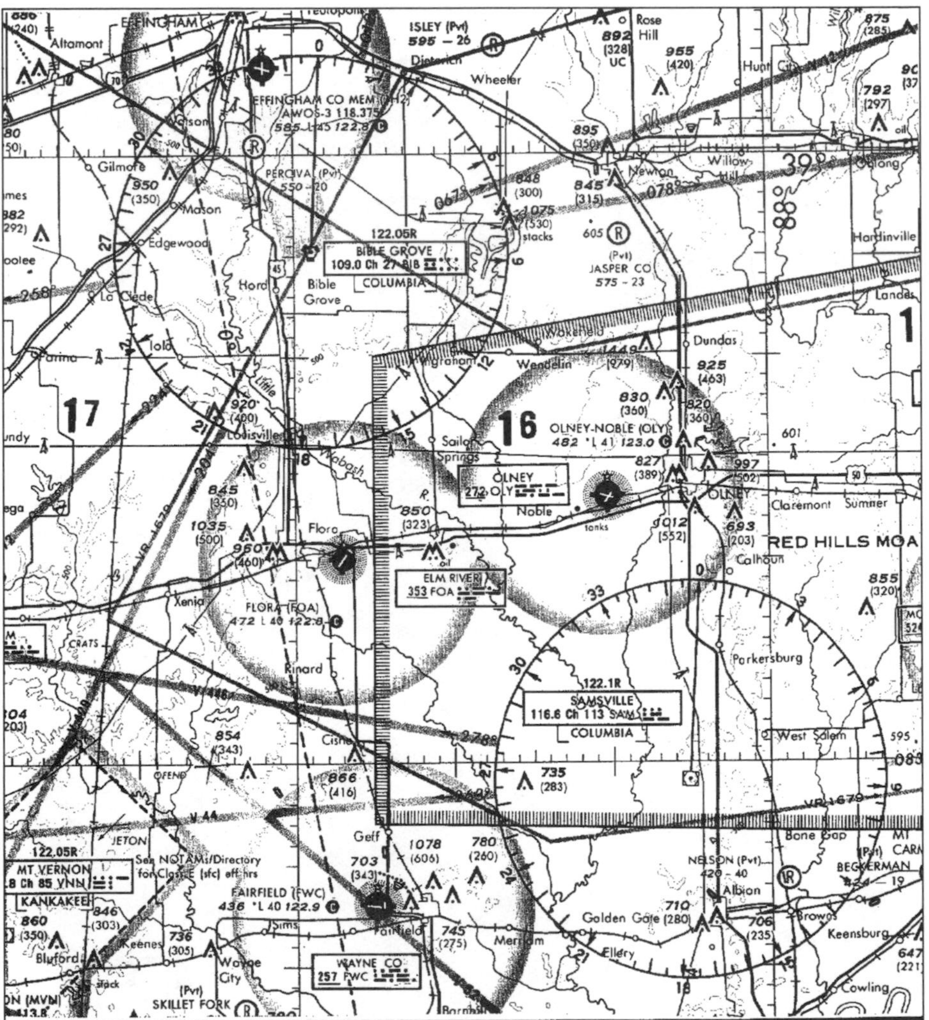
EFFINGHAM
Altamont
ISLEY (Pvt)
595 – 26
Rose Hill
892
(328)
UC
955
(420)
Wheeler
EFFINGHAM CO MEM (1H2)
AWOS-3 118.375
Gilmore
950
(350)
Mason
Edgewood
PERCIVAL (Pvt)
550 – 20
895
(350)
Newton
Willow Hill
848
(300)
845
(315)
1075
(530)
stacks
605
(Pvt)
JASPER CO
575 – 23
122.05R
BIBLE GROVE
109.0 Ch 27 BIB
COLUMBIA
Hord
Bible Grove
La Clede
Iola
Hardinville
Landes
Wakefield
1449
(979)
Dundas
Wendelin
925
(463)
830
(360)
820
(360)
17
920
(400)
Louisville
16
OLNEY-NOBLE (OLY)
482 *L 41 123.0
Sailor Springs
827
(389)
997
(502)
601
OLNEY
272 OLY
Noble
tanks
Olney
Claremont
Sumner
845
(350)
850
(323)
1012
(552)
693
(203)
RED HILLS MOA
1035
(500)
Flora
960
(460)
Calhoun
ELM RIVER
353 FOA
855
(320)
Xenia
FLORA (FOA)
472 L 40 122.8
Parkersburg
Rinard
122.1R
SAMSVILLE
116.6 Ch 113 SAM
COLUMBIA
854
(343)
Cisne
West Salem
595
866
(416)
735
(283)
V 44
Geff
1078
(606)
780
(260)
Bone Gap
122.05R
MT VERNON
KANKAKEE
JETON
See NOTAMs/Directory for Class E (sfc) eff hrs
703
(343)
NELSON (Pvt)
420 – 40
BECKERMAN
FAIRFIELD (FWC)
436 *L 40 122.9
Albion
846
(303)
860
(350)
Keenes
736
(305)
Sims
Fairfield
745
(275)
Merriam
710
Golden Gate (280)
706
(235)
Browns
Keensburg
647
(221)
Bluford
Wayne City
WAYNE CO
257 FWC
Ellery
Cowling
(Pvt)
SKILLET FORK

Fig.18.3

19
NDB introduction

THE NON-DIRECTIONAL BEACON (NDB) IS A RADIO NAVIGATION AID that transmits in the Low to Medium Frequency range. Unlike the VOR system that sends out two signals (the reference and variable phase signals) the NDB sends out only a single signal. This single signal broadcasts from the NDB antenna in every direction, but when the radio in the airplane receives the signal it cannot determine the radio's location. The NDB gets its name from the fact that this radio cannot by itself determine a direction in relation to the station, like a VOR. Therefore it is "non-directional."

The NDB stations broadcast in the same frequency band as AM commercial broadcast stations. The left side of most AM radio dials stops at approximately 510. The NDBs broadcast on frequencies that are farther left or just off the A dial.

Most people these days have had little experience with AM radio. Today the FM band is most popular because of its clarity and sound quality. But if you ever do listen to an AM station, you will recognize two things. First, there is more static on these frequencies because a static discharge, like lightning, will cause interference and the transmission will crackle. Second, the position of the radio will determine the reception quality. If you hold a radio in your hand that is playing an AM station, by turning the radio around, the reception will be affected. In fact, you will find that in certain positions the station will fade out to nothing. This "fade out" is called the "null" position and this is the key to understanding how the NDB radio works.

The first NDB radio for air navigation went into service in 1924. In those days an antenna was mounted on the bottom of airplanes to receive the NDB signal. But this antenna did not look like others. It was not a straight length of wire, instead it looked like a "loop." If you take a common wire clothes hanger and bend its shape into a ring, you will have a "loop" antenna. The loop antenna was connected to a rotating mount. The pilot would tune in an NDB station and then rotate the antenna with a hand crank. Most radio listeners want to turn the radio to get the best reception, but in this case the pilot wanted to find that position that would cause the signal to "fade out." When the plane of the loop became perpendicular to the signal the "null" would occur. When the NDB's signal came through the air and hit both sides of the loop at exactly the same time, the signal on both sides of the loop would be equal and the signals would cancel each other out, thus the fade out. When the loop was not perpendicular to the signal, one side of the loop would get the signal first and strongest while the other side received the signal last and weaker. In this case the signal would not cancel out and the pilot would hear the AM broadcast. When the "null" position was found, the pilot would know that the signal was coming from a direction perpendicular to the plane of the loop. But there are two perpendicular directions: one each way from the plane of the loop. To complete the picture the pilot needed a "sense" antenna to determine which perpendicular pointed the correct direction to the NDB station. So the system needed two antennas, one shaped like a loop, and the other strung across the airplane (usually from the top of the fuselage to the top of the tail referred to as the "clothes line.")

This loop and sense antenna combination was called the Direction Finder (DF). There was no instrument inside the airplane for the pilot to view. This was all done by listening to the radio when tuned to an NDB station. The pilot could determine a position by comparing the direction the signal was coming from and the airplane's heading. The pilot could also "home" to the station by turning the airplane in the direction that the NDB signal was coming from.

As you can imagine, it was a great advancement when a cockpit instrument was invented to display the NDB information on a dial. The modern version of the receiver has a digital tuner for the NDB stations and a dial that points the direction of the station. This modern system is called the Automatic Direction Finder or ADF.

The NDB is the broadcast station on the ground that is received by an ADF in the airplane. Today, the loop antenna is simply a box attached to the underside of the fuselage and the null position is found electronically. The pilot does not have to crank the antenna anymore.

The ADF system (figure 19.1) in the airplane consists of a radio placed into the instrument panel and a Relative Bearing indicator. The Relative Bearing indicator has a Bearing Pointer that will point the position of an NDB station. The

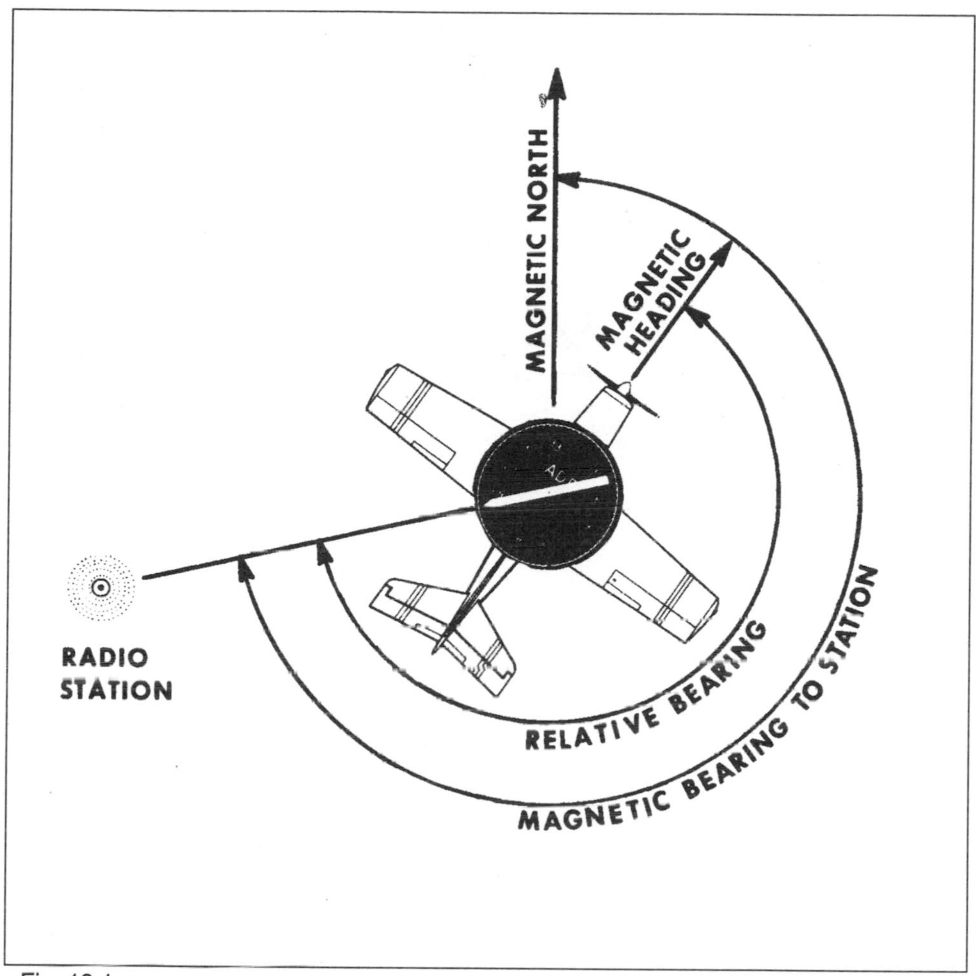

Fig. 19.1

Bearing Pointer sweeps around a dial that looks like a compass because it has all 360 degrees marked off. But this dial is not a compass. It does not indicate the direction that the airplane is traveling. More advanced systems do incorporate a compass and ADF combination called a Radio Magnetic Indicator (RMI), but the basic system does not have this feature. The compass look-alike simply is used to indicate the degrees that an NDB station is located relative to the nose of the airplane.

If a controller ever said: "Traffic is at your 3 o'clock position," where would you look to see the traffic? You would look out past the right wing. Six o'clock

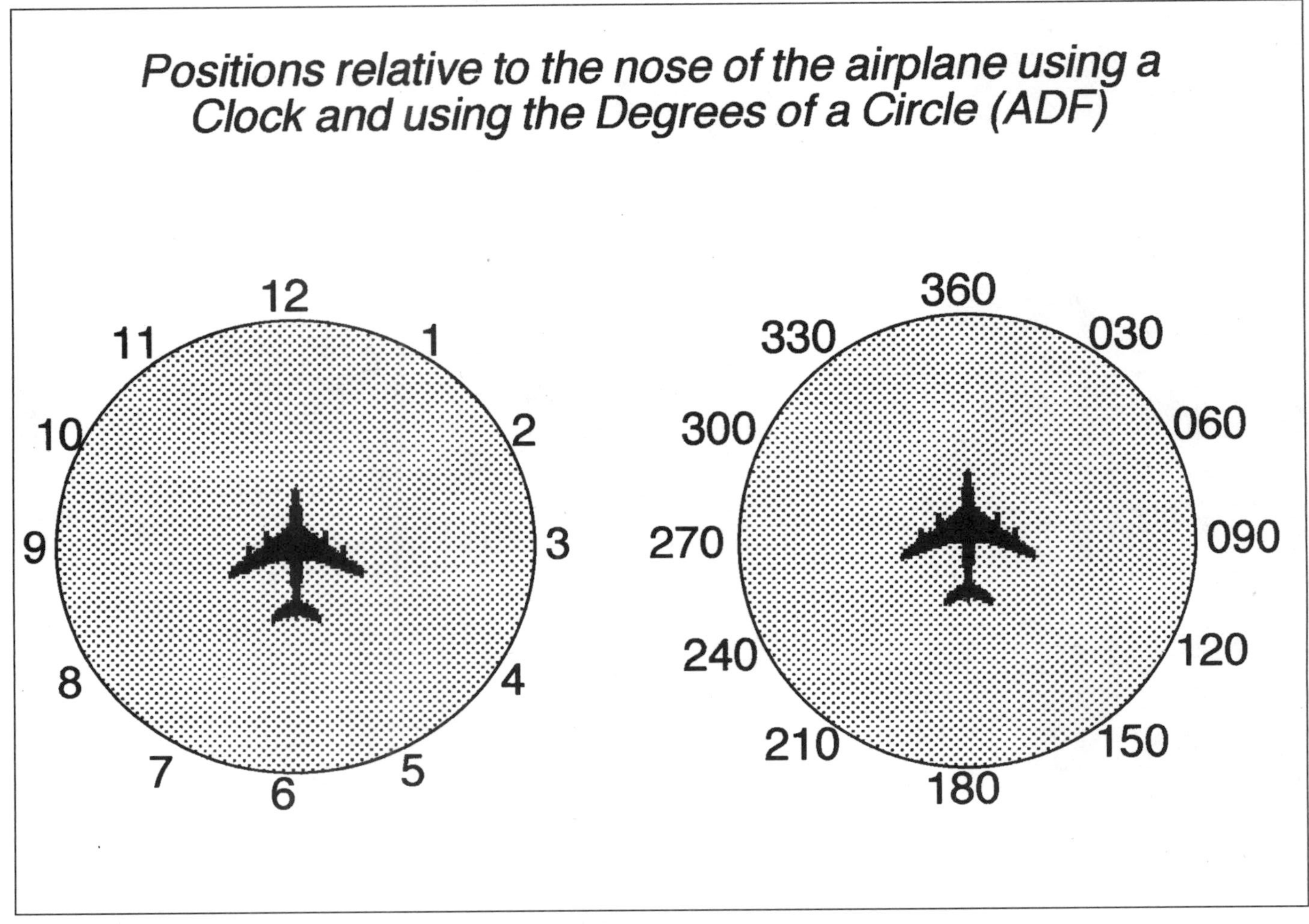
Positions relative to the nose of the airplane using a Clock and using the Degrees of a Circle (ADF)
12
1
2
3
4
5
6
7
8
9
10
11
360
030
060
090
120
150
180
210
240
270
300
330

Fig. 19.2

is the tail position and 9 o'clock is the left wing. Twelve o'clock is dead ahead (figure 19.2). The controller uses a clock face to help pilots get an idea of the position of the traffic. The traffic's position is called out in relationship to your airplane's nose, with 12 o'clock always being straight ahead. The NDB works the same way. The pointer points at the station in relationship to the nose of the airplane. But a clock only has 12 positions and for accurate navigation 12 possible positions is simply not enough. So the ADF system uses the 360° of a circle instead of the 12 positions of a clock. The nose becomes a Relative Bearing of 360°, the right wing is 090°, the tail is 180°, and the left wing is 270°. If the ADF pointer is aiming at 045°, the pilot knows that the NDB station is ahead of the airplane and to the right. The actual location would be approximately halfway between the nose and right wing. If the Relative Bearing (the pointer) was 225°, then the airplane has passed the station and it is behind the airplane and to the left.

WHERE AM I? USING ADF

To determine the heading that a pilot should fly to travel to the NDB station, a little math is used. If you know the Relative Bearing (the indication of the ADF pointer), and you know your current magnetic heading (the indication on a properly set Heading Indicator) you can find the magnetic heading to the station.

The math formula is: Relative Bearing + Magnetic Heading = Magnetic Heading to the Station.

If an airplane is flying East (magnetic heading of 090°) and the ADF reads 045° (Relative Bearing), what direction would take the airplane directly to the station? Use the math formula: 045° (Relative Bearing) + 090° (Heading) = 135° (to the station).

If the pilot were to turn the airplane to fly 135° the ADF pointer would also turn and be aiming at 360° (the nose) when the airplane arrived at a heading of 135°.

Relative Bearing
+ Magnetic Heading
= Magnetic Heading to the Station

What if an airplane is flying South (Heading 180°) and the ADF pointer indicated a Relative Bearing of 270°? What is the heading to the station? $180^{\circ} + 270^{\circ} = 450^{\circ}$. But wait a moment, there is no such heading as 450°. The compass stops at 360° and starts over again past North. To solve this situation remember that anytime the math formula produces a heading that is greater than 360° then 360° must be subtracted to get the correct answer. In this example the answer is $180 + 270 = 450 - 360 = 090^{\circ}$.

Now that you know the math formula, let me say that in real flying you will not use it much. Rather than doing a math problem while trying to fly an airplane, pilots can simply turn the airplane and continue turning until the ADF pointer is aiming at the nose. Then the pilot can read what heading will place the pointer on the nose. All this without math! The math formula is presented here to help pilots get a foundation of information.

If I am flying a heading of 270° and the Relative Bearing is 180°, I do not need a math formula to know that I am flying West and the NDB station is right behind me. I have passed the station. If I want to fly over the station I must turn around. It is very important to "picture" the location of the NDB station in your imagination when flying. Is the station up ahead and to the left (Relative Bearing of 330°)? Or is it behind and to the right (Relative Bearing of 165°)? In each case the pilot must "see" the location of the station and the position of the airplane with regard to the station.

Today most NDB stations are low power facilities. The long range navigation has been taken over by the VORs. Once there were NDB's that had a range of 200 miles, but the FAA began to decommission these in 1965. The routes that were once made up by NDBs were "color" coded. Red and Green Routes generally ran North and South. Blue and Yellow Routes ran East and West. Together these made up the "colored airways." There are very few of these colored routes left, because there are very few long range NDBs remaining. NDBs used today are used for instrument approaches to a particular airport and not for enroute navigation. With that said, do not think that a knowledge of NDBs is not important. A basic understanding now will help greatly later in your instrument flight training.

What is the toughest instrument approach to fly? It is not the ILS. No, the toughest approach is an NDB approach in a crosswind!

Verify these NDB practice problems:

Magnetic Heading (Heading Indicator)	Relative Bearing (ADF pointer)	Magnetic Heading to the station
1. 180°	090°	270°
2. 080°	360°	080°
3. 010°	180°	190°
4. 270°	270°	090°
5. 045°	045°	090°
6. 140°	180°	320°
7. 220°	270°	130°
8. 330°	045°	015°
9. 085°	315°	040°
10. 255°	225°	120°

20 ADF indications and positions

IN CHAPTER 19 THE BASICS OF THE NDB SYSTEM WERE DISCUSSED, BUT this chapter will show how the system is actually used to determine positions in the airplane. The sample problems at the end of the last chapter required a math calculation, but as stated before, there really is not much time in flight to do math problems. The NDB system must become a "mental" exercise. The pilot must picture the airplane's position in relation to the station by evaluating the heading and relative bearing.

If the airplane is flying with a heading of 270° and the ADF indicator shows a relative bearing of 315°, where is the airplane in relation to the station? South? Northwest? East? The math problem would be 270 + 315 = 585 - 360 = 225. This means that the pilot would have to fly a heading of 225° to fly to the NDB station. An indication of 225° is a Southwest direction. If the pilot must fly Southwest to arrive at the station that must mean that the airplane is already Northeast of the station. Figure 20.1 illustrates this situation. This figure is a view from above looking down on the NDB station and on the airplane. This is the view that the pilot must formulate in the imagination to understand the position. The airplane is Northeast of the station and must now fly Southwest (225°) to fly to the station.

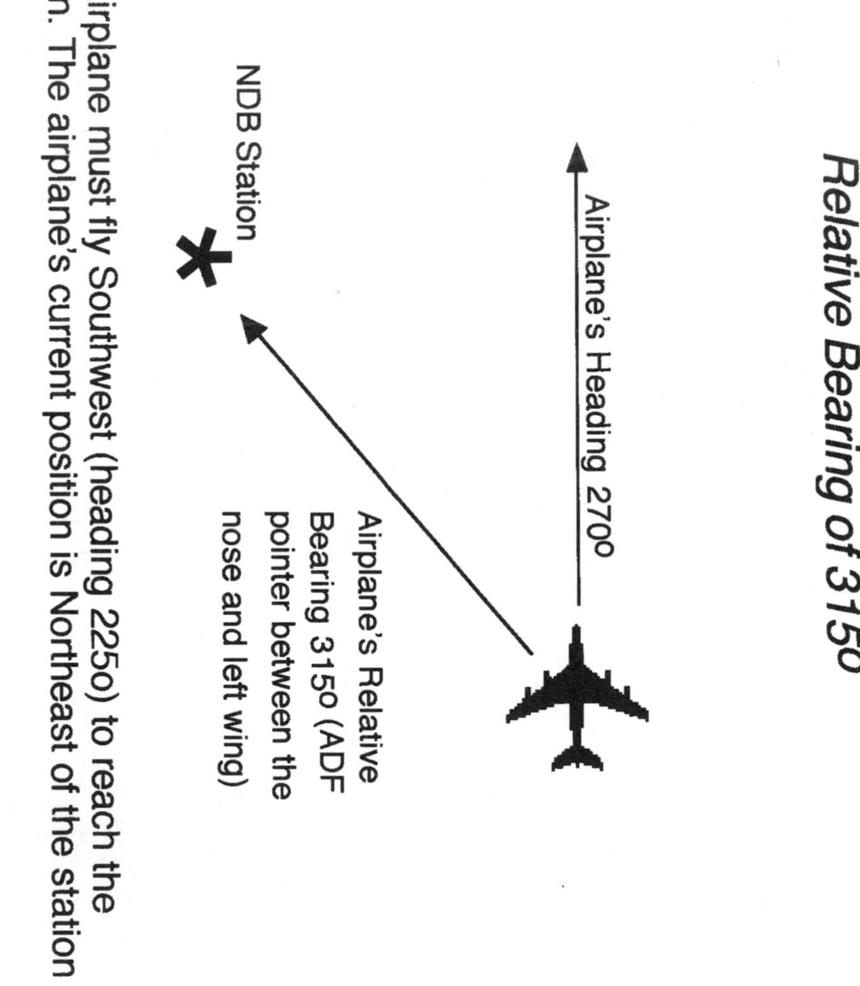
Position of airplane with a Heading of 270o and a Relative Bearing of 315o
Airplane's Heading 270o
Airplane's Relative Bearing 315o (ADF pointer between the nose and left wing)
NDB Station
The airplane must fly Southwest (heading 225o) to reach the station. The airplane's current position is Northeast of the station

Fig. 20.1

21
LORAN and GPS

THE TERM LORAN IS AN ACRONYM FOR LONG RANGE NAVIGATION. LORAN is actually a maritime system used by large cargo ships and small boats alike. The LORAN system provides coverage primarily over the North Atlantic and the North Pacific Oceans, where navigation aids for airplanes are not in place. LORAN has been adapted to airplane use and is providing a long range navigation system that is reasonable in cost. LORAN distinguishes itself from other forms of navigation in aviation in that it is hyperbolic navigation, or rho-rho navigation (rho-rho navigation uses distances exclusively). The VOR-TAC system uses the rho-theta system of navigation. Because of the different uses and applicability in using LORAN, several modifications to the system have taken place. The improvements to the LORAN system have caused it to be identified as A, B, C, and D.

BACKGROUND

LORAN-A was intended mainly for maritime operations. This system operated in the medium frequency band just above the AM broadcast band. This frequency is in the 1700-2000 kHz band. This system consists of a master station and a secondary station (originally called a slave station), which are located about 500 nautical miles apart. The master station sends out a signal at precise intervals. The secondary station receives this signal and transmits another coded signal on the same frequency. This gap in transmission causes the LORAN receiver on board the vessel to receive different signals on the same frequency. The receiver measures the time delay and displays this information to the operator. This time delay information can be used to plot a line of position (LOP) where the vessel is located. This entire procedure is used with a second pair of stations. This second LOP will cross the first LOP and define the vessel's position. Each LOP must be plotted on a graph by an operator in order to determine the crossing point of the two LOPs. The disadvantage of using this system is the time involved. When determining the first LOP, the vessel is leaving the first LOP. Since an airplane's speed is much greater than that of a boat, determining an exact position for the airplane is difficult, if not impossible. The LORAN-A system was operated by the United States Coast Guard and has been decommissioned.

LORAN-C is a sophisticated system of navigation that encompasses master stations and secondary stations. This system is known as a long range system of area navigation. The signals that are transmitted by the master and secondary stations generally follow the contour of the earth's surface and can be received from great distances. This operating characteristic of LORAN-C allows the user distances much greater than line of sight distance which is associated with VHF. However, LORAN-C can be used over any distance, whether it is long or short. By using the system over long distances, the advantages can be seen over VOR navigation, which does not have long range capabilities. The short range systems also treat the earth as a flat surface, not curved. Traveling over short distances, this is not a problem. However, covering long distances, substantial distance and bearing errors would occur if the navigation system does not encompass Great Circle Routes.

THEORY

LORAN-C is the current civilian version of the LORAN system and is similar to LORAN-A. The major difference between the two systems is that the LORAN-C receiver uses a microprocessor to determine quickly and accurately its position from the LOPs received. The Loran LOPs are not straight lines, but hyperbolas (figure 21.1). This is a function of how LORAN-C operates.

The LORAN-C system requires a network of transmitting stations referred to as a chain, to determine a position. One of these stations is referred to as the master station, the others (two or more) are called secondary stations. The master station is designated as an M, while the secondary stations are designated V, W, X, Y, and Z. This combination of the master station and two or more of the secondary stations is known as the chain. There are 17 LORAN-C chains located throughout the world, with nine located in the United States. At precise intervals the master station transmits a signal in the Low Frequency band of 100 kHz. These signals are sent out at intervals with certain times between each signal. This time interval identifies the chain of stations as the Group Repetition Interval (GRI). The GRI of 89,700 milliseconds identifies the Great Lakes Chain as the GRI-8970 chain (figure 21.2), which is located in the midwest.

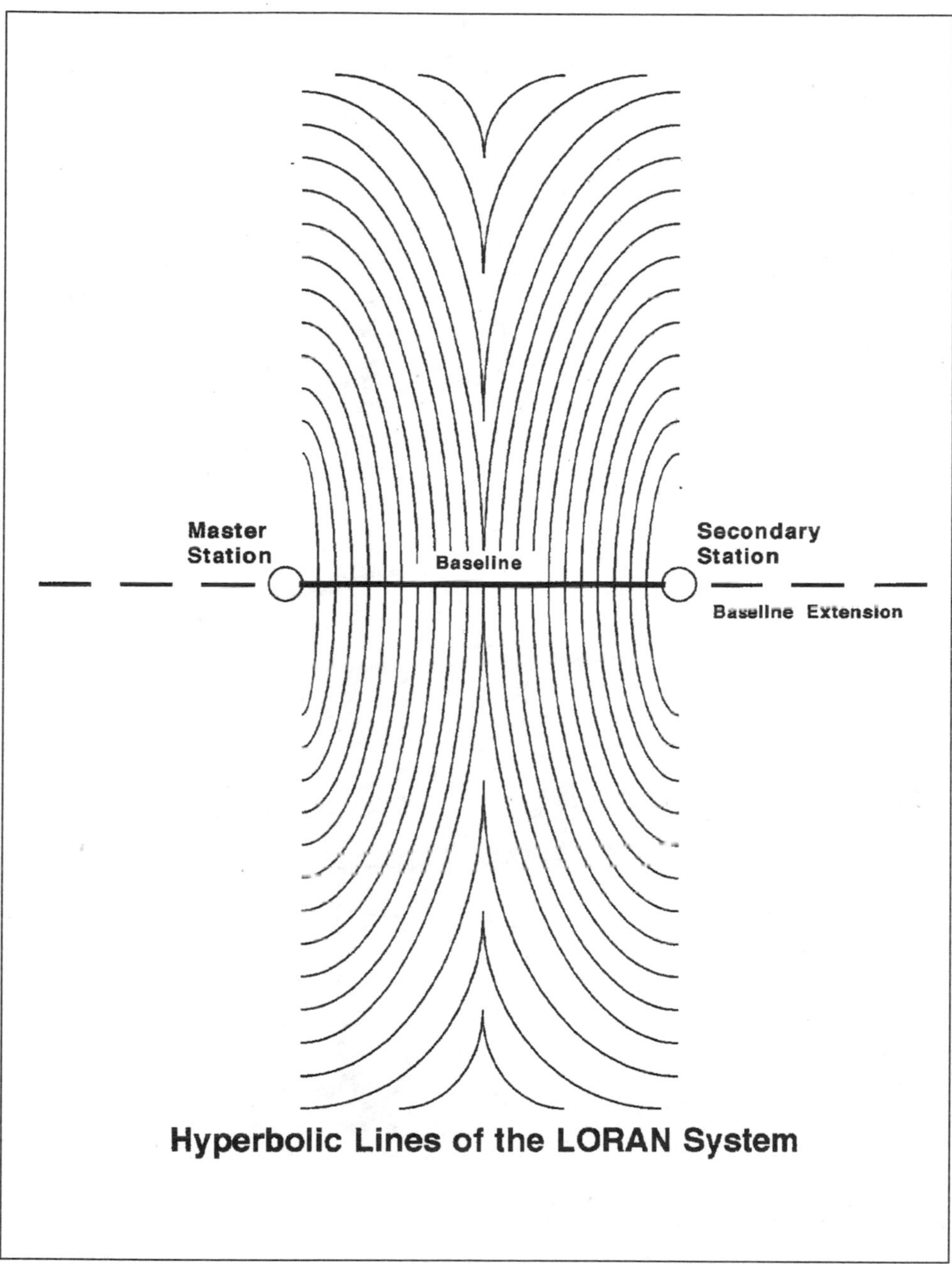
Master
Station
Baseline
Secondary
Station
Baseline Extension
Hyperbolic Lines of the LORAN System

Fig. 21.1

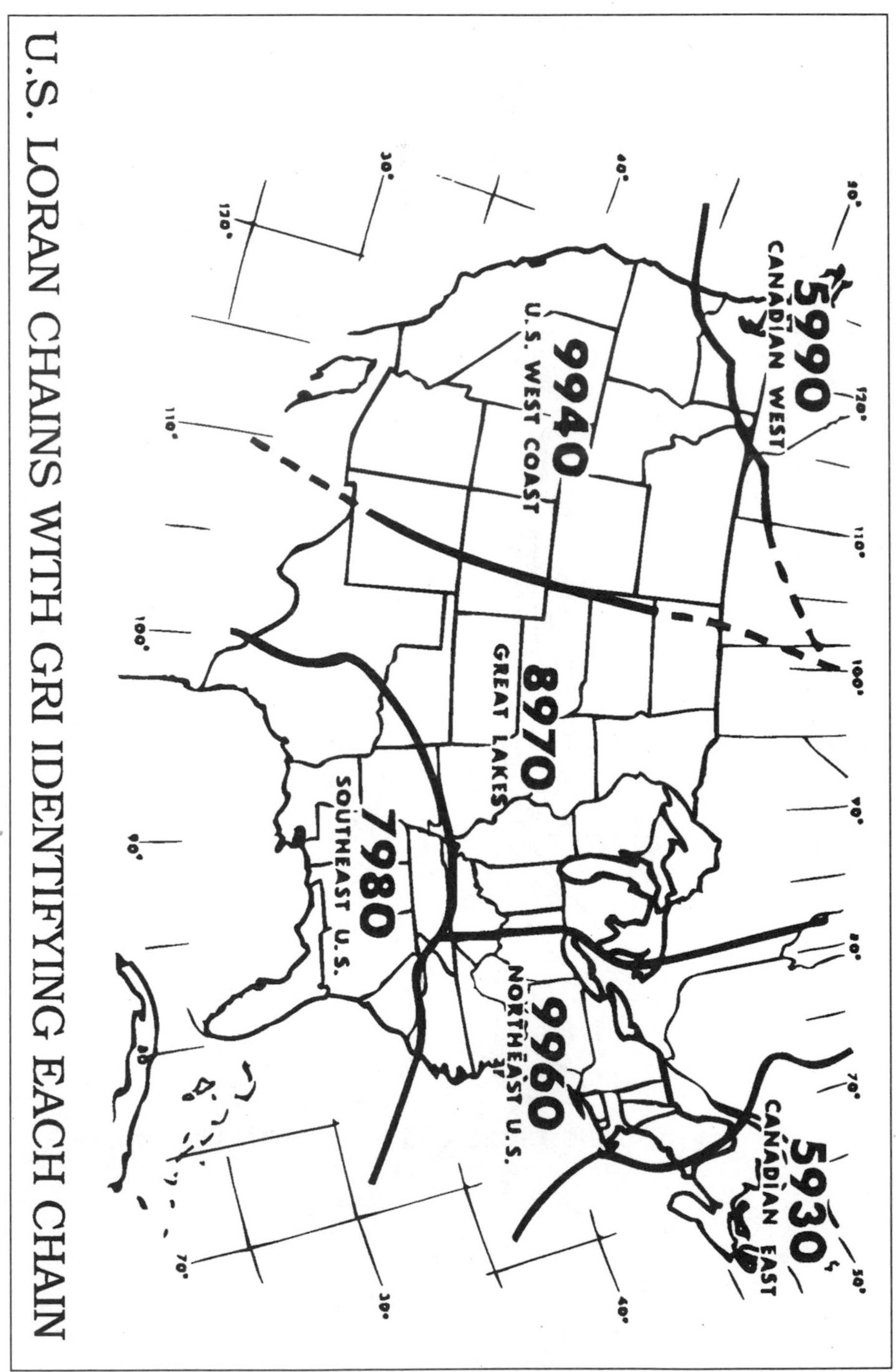
U.S. LORAN CHAINS WITH GRI IDENTIFYING EACH CHAIN
5990
CANADIAN WEST
9940
U.S. WEST COAST
8970
GREAT LAKES
7980
SOUTHEAST U.S.
9960
NORTHEAST U.S.
5930
CANADIAN EAST

Fig. 21.2

The master station sends out a string of signals that in turn causes the secondary stations to send out similar signals. The airplane receiver decodes all of these signals, identifies the chain that is being received, and measures the time differences between the master station and each of the secondary stations received. The LORAN receiver uses the time differences to plot LOPs. By using two or more secondary stations, two or more LOPs are plotted, thus allowing the receiver to display the aircraft's location in terms of latitude and longitude.

The hyperbolic LOPs are the key to the operation of the LORAN system. The LORAN-C uses the time difference in the receipt of the master and secondary signals. The receipt of the time difference is transmitted into a hyperbola. The airplane can be anywhere on the hyperbolic LOP because of the principle involved. When the object (in our case the airplane) is plotted on the hyperbolic line, there is always the same difference in distance anywhere along that line (figure 21.3). By looking at figure 21.3, note that two hyperbolas are possible with the same difference in distance. The LORAN-C receiver determines which signal was received first, the master or the secondary, and then determines which hyperbola is the correct one. The second hyperbola is then eliminated. The result is a single hyperbolic LOP for each particular pair of master/secondary station signals. By using the hyperbolic LOP from the first calculation, then using the same master station and a different secondary station, an intersection of the two hyperbolas define the aircraft's position (figure 21.4).

The present position feature of the LORAN system is only one of the many uses of the system. The LORAN computer displays course guidance and distance information, groundspeed, ground track, range, distance to go and estimated time of arrival, as well as other features that can be purchased depending upon the unit installed. These LORAN-C features are all advancements over previous LORAN systems, which required all LOPs to be plotted, and did not provide any additional information beyond the LOP.

ACCURACY

The accuracy of the LORAN-C can vary much in the same way VOR and NDB accuracy can vary. The accuracy of the system is in direct proportion to the distance from the transmitting station. However, it should be noted that LORAN-C is still much more accurate than VOR navigation. LORAN-C accuracy is most often measured in two ways, repeatable accuracy and absolute accuracy.

Repeatable accuracy refers to returning to a specific location with guidance to that location. This is the ability to return to that same location using the LORAN's previously obtained coordinates. Put another way, repeatable accuracy is the ability to find a position where the unit was previously. An ideal signal reception can result in repeatable accuracy within 50 feet. The average repeat-

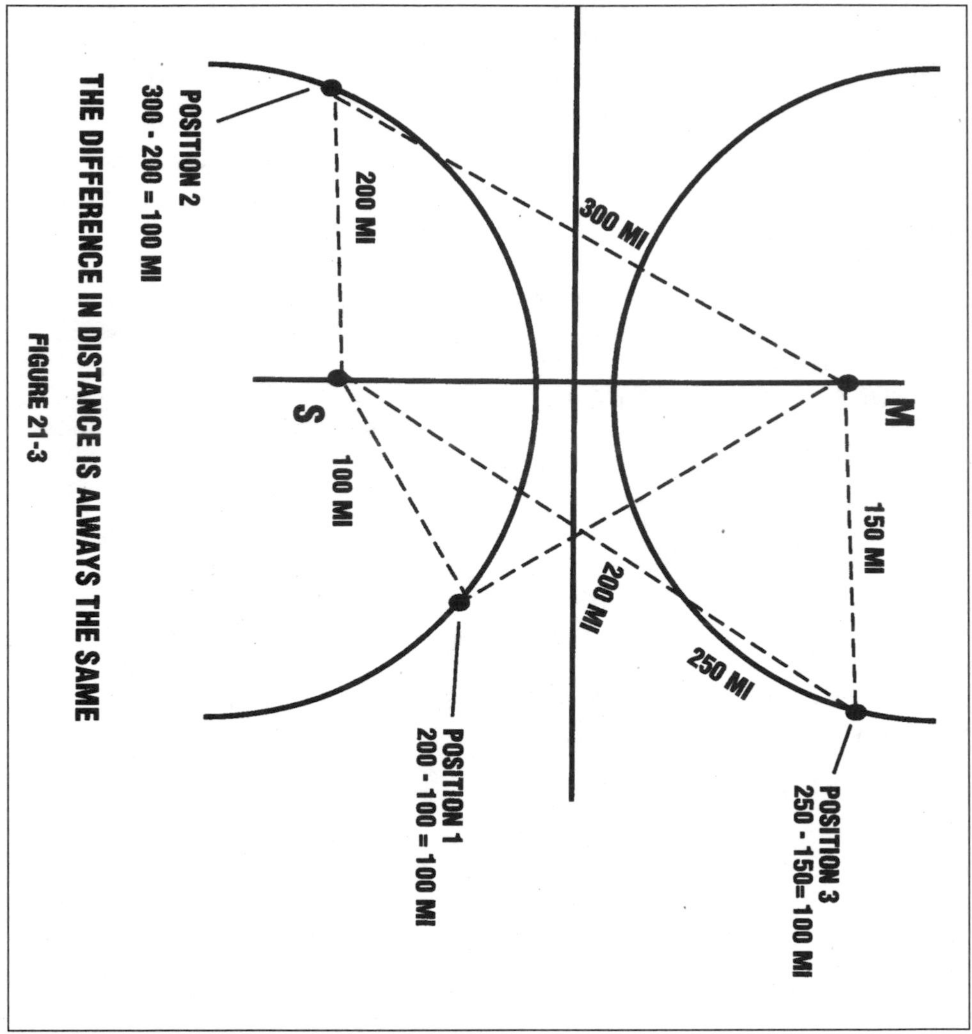

Fig. 21.3

able accuracy is generally about .01 nautical mile, or around 60 feet. The reason that the LORAN unit can be this accurate for a return is because terrain and propagation errors are about the same for the return, thus eliminating those errors.

Absolute accuracy refers to the ability to determine present position independently. Put another way, absolute accuracy is the accuracy of the latitude and longitude computed by the unit, compared to the latitude and longitude on the chart. This accuracy can range from .1 nautical mile to 2.5 nautical miles.

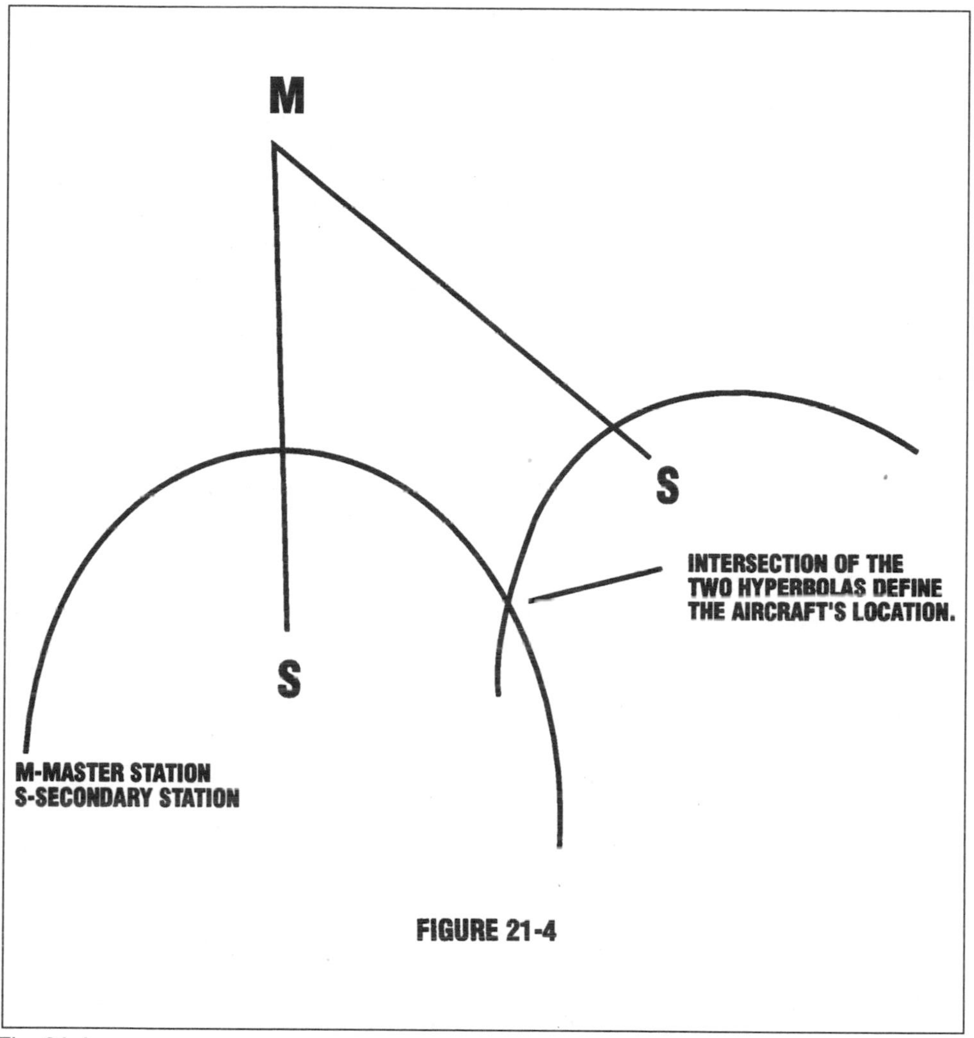

Fig. 21.4

LIMITATIONS AND DISADVANTAGES

LORAN-C is a navigation system that is outside the Federal Aviation Administration's navaid system. There are some limitations that restrict LORAN-C navigation. Since the original LORAN was used for maritime use, the orientation toward overwater areas is apparent. This becomes a factor when the overall coverage of the LORAN system is looked at as a whole. Other areas of concern that have an effect on the LORAN system are the presence of ground waves and sky waves. One other factor that is involved is the problem of precipitation static and its effects on the LORAN unit in the airplane.

The overwater orientation of the original LORAN system still has some application to today's LORAN system. The assumptions of the original system regarding signal speed and transmission were all based on movement over open seawater. LORAN signal propagation is approximately six microseconds per nautical mile over open seawater. When the signal is over land, the terrain causes the signal speed to become less and variable. As a result, the signals over land will be slightly in error. This effect is known as additional secondary factor (ASF), and is corrected for by the LORAN-C unit. Even though those errors are not usually significant, they can affect the accuracy to some extent if not corrected. The maximum accuracy is thus obtained when flying over open seawater using coastal LORAN chains.

The coverage of the LORAN chains is a limiting factor of the system (figure 21.5). The LORAN-C system covers the continental United States, its coastal waters, and various parts of Alaska (with emphasis on the Alaska coastline). Complete coverage of the continental United States was completed in May of 1991, thus eliminating what was known as the "mid-continent gap." Since the system is generally maritime based, the North Atlantic and the North Pacific Oceans are fairly well covered. However, other areas of the world lack coverage. There is very little coverage of Mexico, Europe, and the Caribbean. There is also no coverage in South America or Australia, with limited coverage in North Africa. Because of this limited worldwide coverage, LORAN cannot be used for worldwide international navigation. The airplanes that fly worldwide must have a different navigation system available.

LORAN-C is also affected by a phenomenon known as sky waves. The LORAN unit prefers the ground wave, which is transmitted and follows the contour of the earth. This type of wave is known as a ground wave and is the most accurate for the unit to use. The sky wave leaves the contour of the earth and is refracted off of the ionosphere. From the ionosphere, the sky wave is then transmitted to the unit. This extra travel of the signal takes a longer period of time and thus affects the plotted LOP. Because of this extra time involved, the accuracy of the LOP is affected. At distances of less than 1,000 nautical miles from the transmitting station, the receiver can differentiate between a ground wave and a skywave. At greater distances the unit must compensate for the resultant ground and sky waves, thus affecting accuracy slightly.

Precipitation static is an electrical charge that builds on an aircraft that results in poor radio reception. This static forms when flying through precipitation and is most noticeable as static from the communications radio. All electronic systems on the airplane are affected to a certain extent. The most common method of dissipating this static is through static wicks in the airstream on the trailing edges of the flight control surfaces. The static that forms is known as noise.

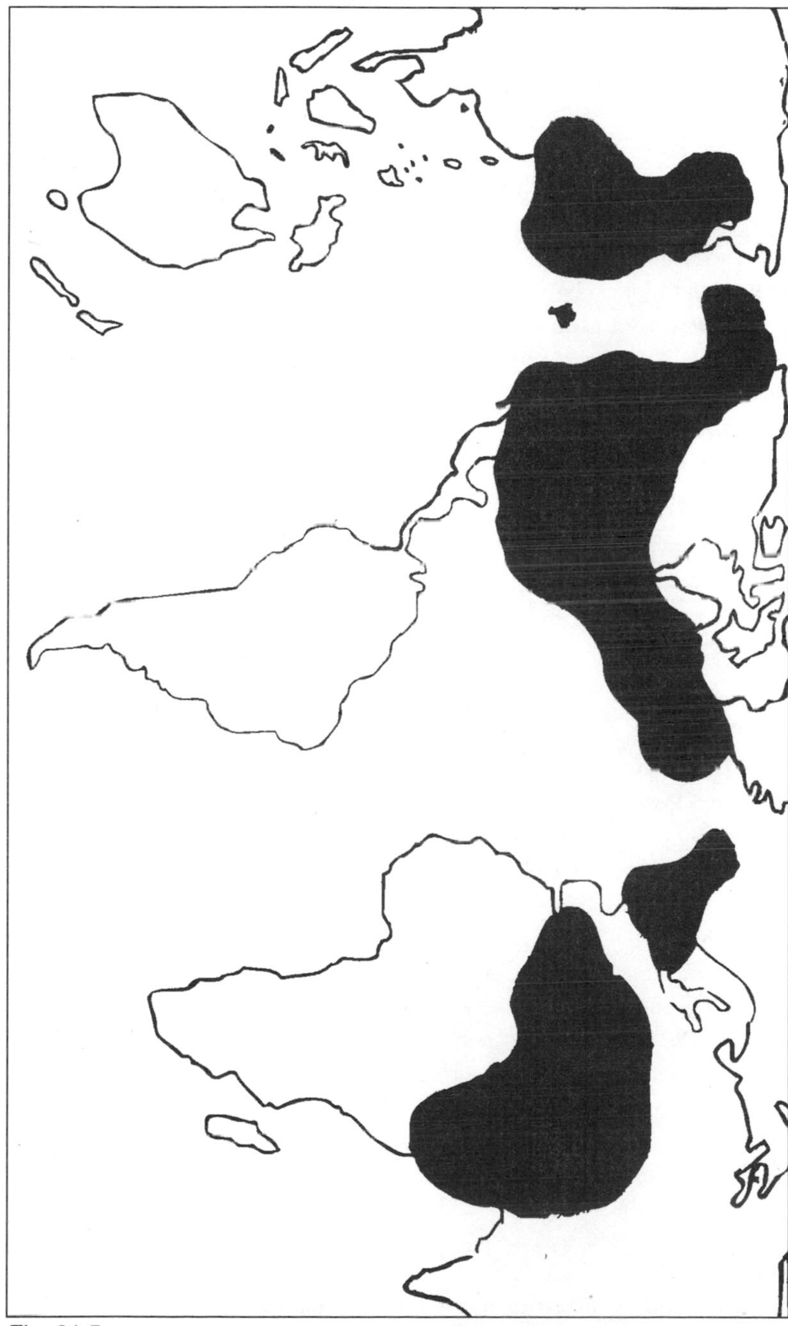

Fig. 21.5

Most LORAN-C receivers have a signal-to-noise ratio (SNR) monitor that alerts the pilot when the signal ratio falls below an acceptable level. SNR is a ratio, which can be caused by the low signal strength, or a high noise level due to static, or a combination of both factors. In order to ensure adequate signal strength, the pilot must leave the area of precipitation if the SNR level falls below an acceptable level.

ADVANTAGES

There are many advantages of LORAN-C navigation. This type of navigation provides a greater usable range than VOR signals. The accuracy of the system is much more accurate than VOR or NDB navigation. The approaches that are available with the LORAN-C system are non-precision. However, the ability to achieve the lowest non-precision approach minimums is greatly enhanced by the accuracy of the LORAN system. Another advantage is that LORAN-C approaches do not require the installation of any additional ground equipment. Perhaps the biggest advantage is that a typical LORAN-C unit is affordable for the general aviation pilot. The average LORAN units are available from $3,000 to $5,000.

NAVSTAR/GPS

The NAVSTAR Global Positioning System is a space based satellite network that provides the user with very accurate three dimensional position information. The Global Positioning System (GPS) will likely be the most important air navigation development since the advent of VOR technology. The GPS provides an accurate three-dimensional position anywhere in the world anytime, day or night. The limitations that are associated with other navigation aids, VHF, VOR, and LORAN navigation, do not apply to GPS. GPS will function virtually free of all environmental limitations. GPS will not only streamline air navigation during the enroute phase, it will also assist in approach navigation as well. As the GPS technology is perfected, Category I, II, and III precision approaches will most likely become a reality. The GPS will most likely supplant less accurate navigation systems such as VOR, DME, LORAN-C and Omega over the coming years. Initial operating capability was realized in late 1993, with GPS satellites operating in their assigned orbits. Full operational capability is scheduled to take place in 1995.

BACKGROUND

GPS technology has evolved in phases. In December of 1973, the military received approval to begin the concept validation phase of the GPS program.

This program included system performance studies as well as feasibility studies. This phase was completed in 1979. The second phase was begun after the completion of Phase One and included the development of the GPS user equipment and testing. Phase Two was completed in 1985. The third phase was started in 1985. This phase included production of GPS equipment and developments that led to the completed satellite group (constellation) and the Master Control Station. In 1986, the overall operation of the Control and Space Segment of the GPS was handled by the USAF 2nd Space Wing at Falcon AFB, Colorado. This facility continues to oversee the overall operation of the GPS.

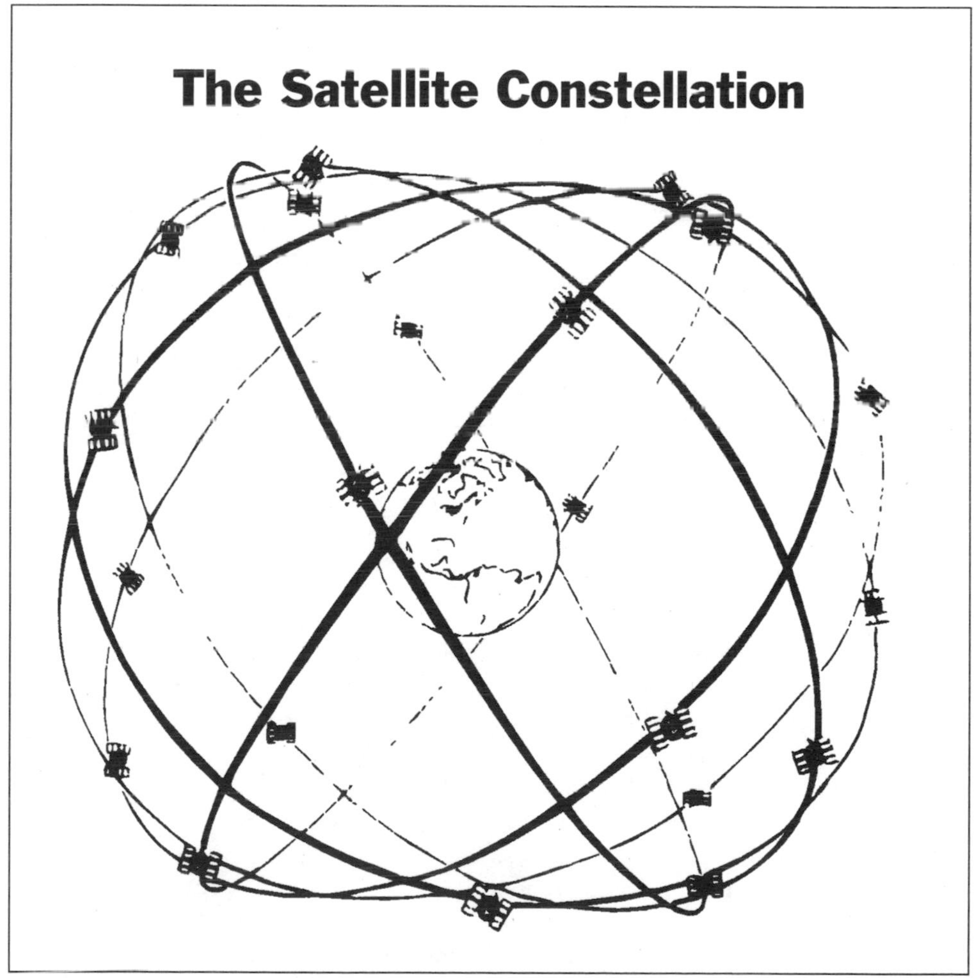

Fig. 21.6

DESCRIPTION

The GPS is a space based system consisting of a group of satellites evenly spaced around the world. There are 24 deployed satellites (three are spares), orbiting at an altitude of 10,900 nautical miles above the earth (figure 21.6). This relatively high altitude allows the satellites to be difficult to destroy and allows worldwide coverage with a fewer number of satellites. The satellites usually complete an orbit in 12 hours. At this rate they will orbit the earth twice each day. The satellites are positioned in such a way that a minimum of four satellites are in view by a user anywhere in the world at anytime. A minimum of four satellites is required for a position fix.

All of the satellites are controlled by a control segment on the ground. The control segment consists of a master control station and five monitor stations. The master control station is located at Falcon AFB in Colorado. The monitor stations are located at Hawaii, Kwajalein, Diego Garcia, Ascension, and a station collocated with the master control station in Colorado (figure 21.7). The master control station is the headquarters for the entire network, and is operational 24 hours a day, 365 days a year. This station is responsible for tracking, managing, and monitoring all of the GPS satellites. The monitor stations are responsible for tracking all GPS satellites that are in view and collect data from each satellite. The monitor stations then send the data to the master control station for use. Updated information is transmitted to each satellite by means of ground antennas. The information sent to the satellites via the control segment corrects satellite position, and recalibrates the atomic clock if necessary. This information ensures the exact position of each satellite at a given time, with that time being very precise. Exact location and exact time are critical for ensuring proper positioning information for the user.

THEORY

GPS is a passive system. The system works all of the time and does not require any interrogation from users. Because the system is passive, there can be an unlimited number of users. The main problem with a passive system is that a very precise time measurement must be involved in order to time the interval from the beginning of the signal to the receipt of the signal by the user. With a traditional active system, a simple time measurement can be used because the user initiated the signal. Therefore the time is known by the user. In the case of GPS, the signals come continuously. Therefore, unless both units at each end (satellite and user) are exactly synchronized, there will be discrepancies in the actual position (a billionth of a second error produces a position error of one foot). Since having exact synchronized clocks on each end is impractical, GPS accepts the fact that the time on the user end will be somewhat inaccurate. By using a fourth satellite to eliminate the time bias, the time error can be eliminated.

Location of Master Control Stations and Monitor Stations

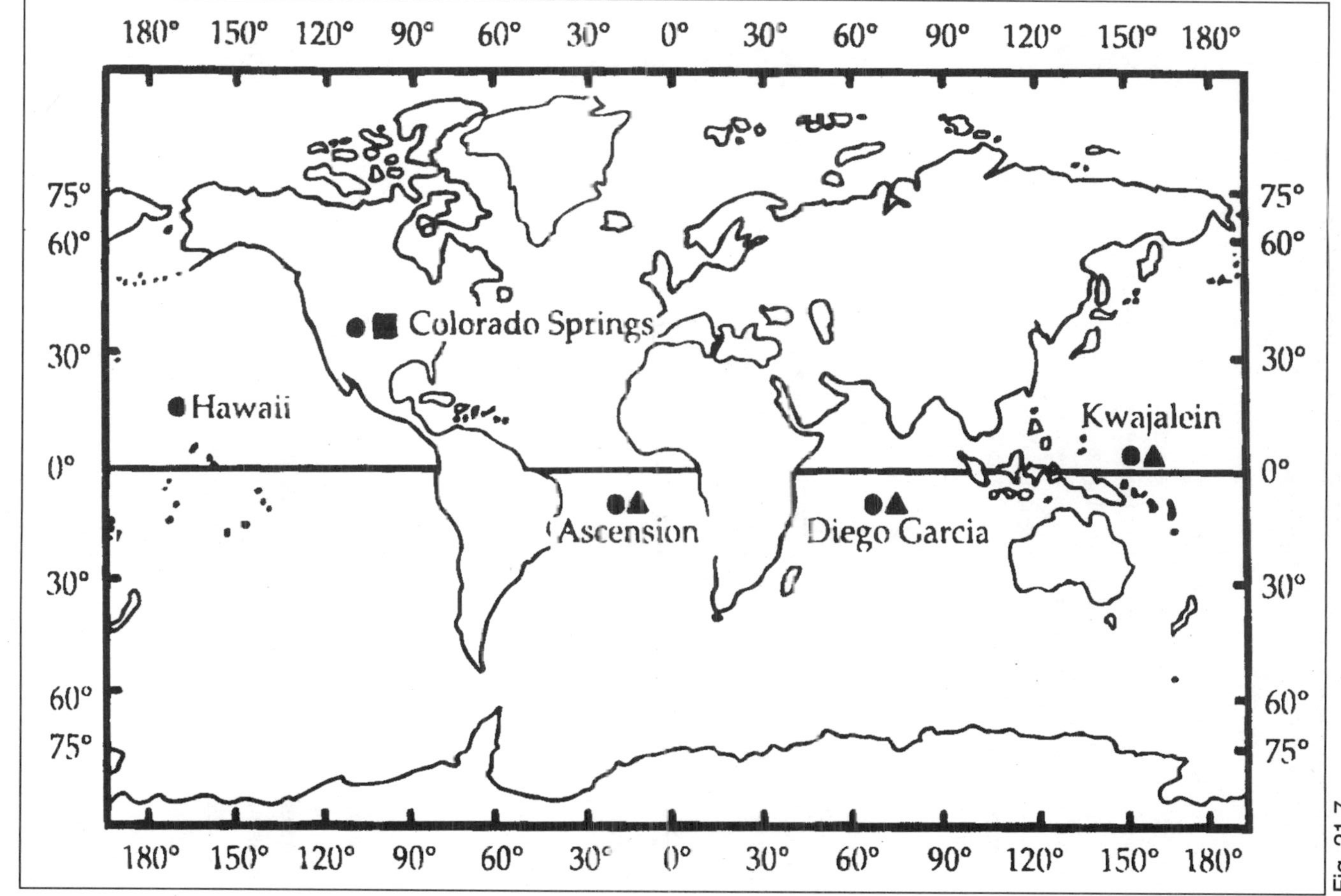

Fig. 21.7

The GPS satellites are equipped with a time device known as an atomic clock. It is not practical to adjust the receiver's clock to zero the time bias. As a result, the accuracy on board the satellite is maintained to an accuracy of a few billionths of a second. The clock stability of the GPS is 10(-13), or .003 seconds per one thousand years. This accuracy is required for the three-dimensional positioning system.

GPS satellites transmit signals on two different frequencies and in two modes. These strings of data bits are generated and timed by an onboard clock that provides the exact time of the broadcast signal. The signals are transmitted at the speed of light and consist of a course/acquisition (C/A) mode, and a precision (P) mode. The P mode is used for the military and is very complex. It is transmitted at a rate of 10,230,000 bits per second and takes a full week to repeat itself. The C/A mode is transmitted at a rate of 1,023,000 bits per second and can be deciphered relatively easily. This code repeats itself every second and can be acquired quickly by the receiver.

Once the GPS receiver has acquired the code and synchronized itself as much as possible with the satellite code, it can then measure the time between incoming codes. The more time difference between the codes, the greater the distance the receiver is from the satellite. This information is gathered from at least three satellites, and multiplies the time difference by the speed of light to produce psuedorange measurements. Each of these measurements is a range from the satellite. The location of the receiver is located on a sphere (ring) that is centered around the satellite (figure 21.8). As the receiver receives input from different satellites, overlapping spheres create a point where the receiver is located (figure 21.9). However, since the satellites are moving at an approximate speed of 7,500 miles per hour, the exact time of the signal must be known. The error in the measurement of the elapsed time between signal transmission and reception is caused by the inaccuracies of the receiver's clock. This time bias error causes distance measurements to be somewhat inaccurate. A fourth satellite is used to correct the time bias, which in effect is replacing the atomic clock that is missing on the receiving end. In effect, three satellites provide spheres of reference, and the other satellite corrects for the time error. That is the reason for the requirement of having four satellites in view for an accurate position. The GPS receiver does all of the calculations automatically. The calculations are then put into a longitude, latitude, and altitude format. All other forms of navigation computations can then be performed by the GPS receiver after the current position is established.

ACCURACY

The accuracy of the GPS can be defined in two ways. The precision positioning service (PPS) is a highly accurate positioning, velocity, and timing service that

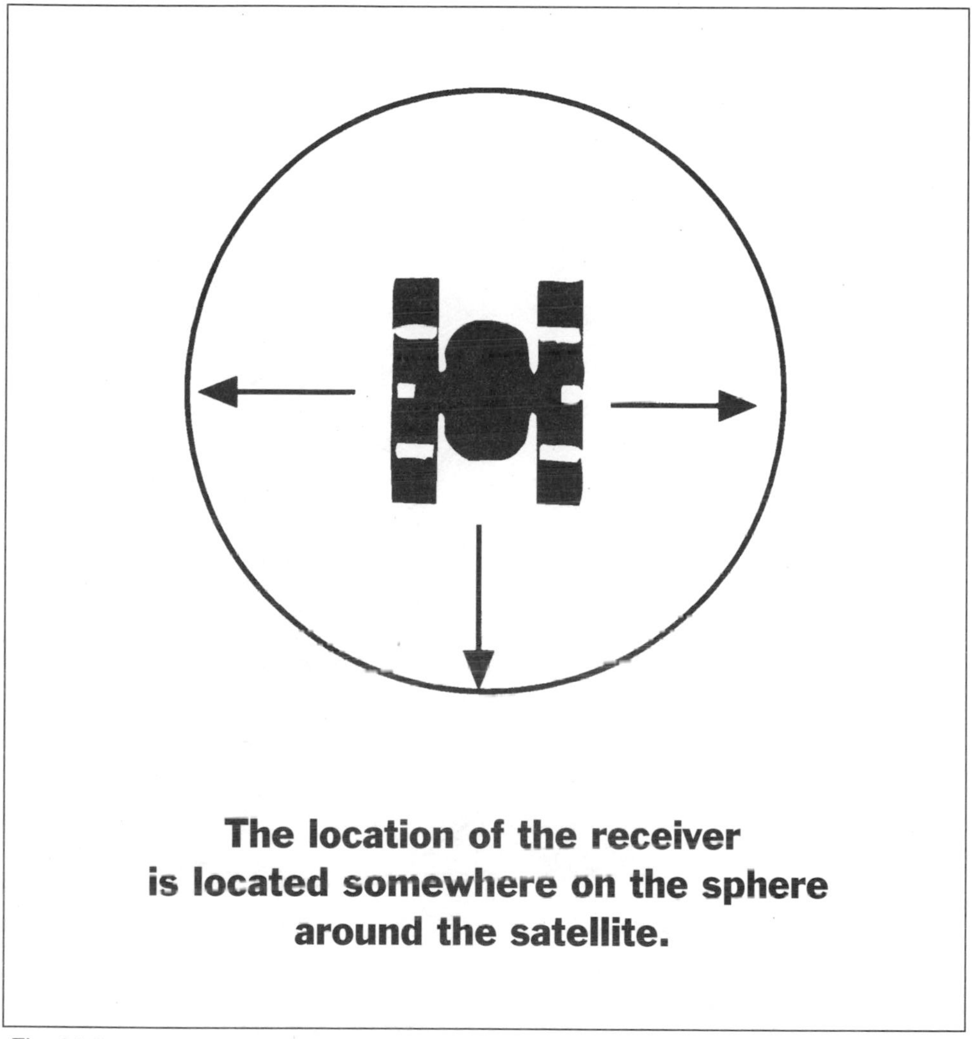

The location of the receiver
is located somewhere on the sphere
around the satellite.

Fig. 21.8

is available only to authorized users (military). The accuracy of the PPS is specified to provide an accuracy of 16 meters (52.49 feet). The military is able to further degrade the GPS position through selective availability (SA). This feature of the system causes the GPS system to be less accurate than the PPS. This less accurate feature is used by the civilian sector. The standard positioning service (SPS) is specified to provide 100 meter (328.1 feet) position accuracy with a 99.99 percent probability (SPS is for civilian use). In reality however, much more accuracy than 328.1 feet is obtained. This type of accuracy can provide enroute (area navigation) navigation and non-precision approach guidance.

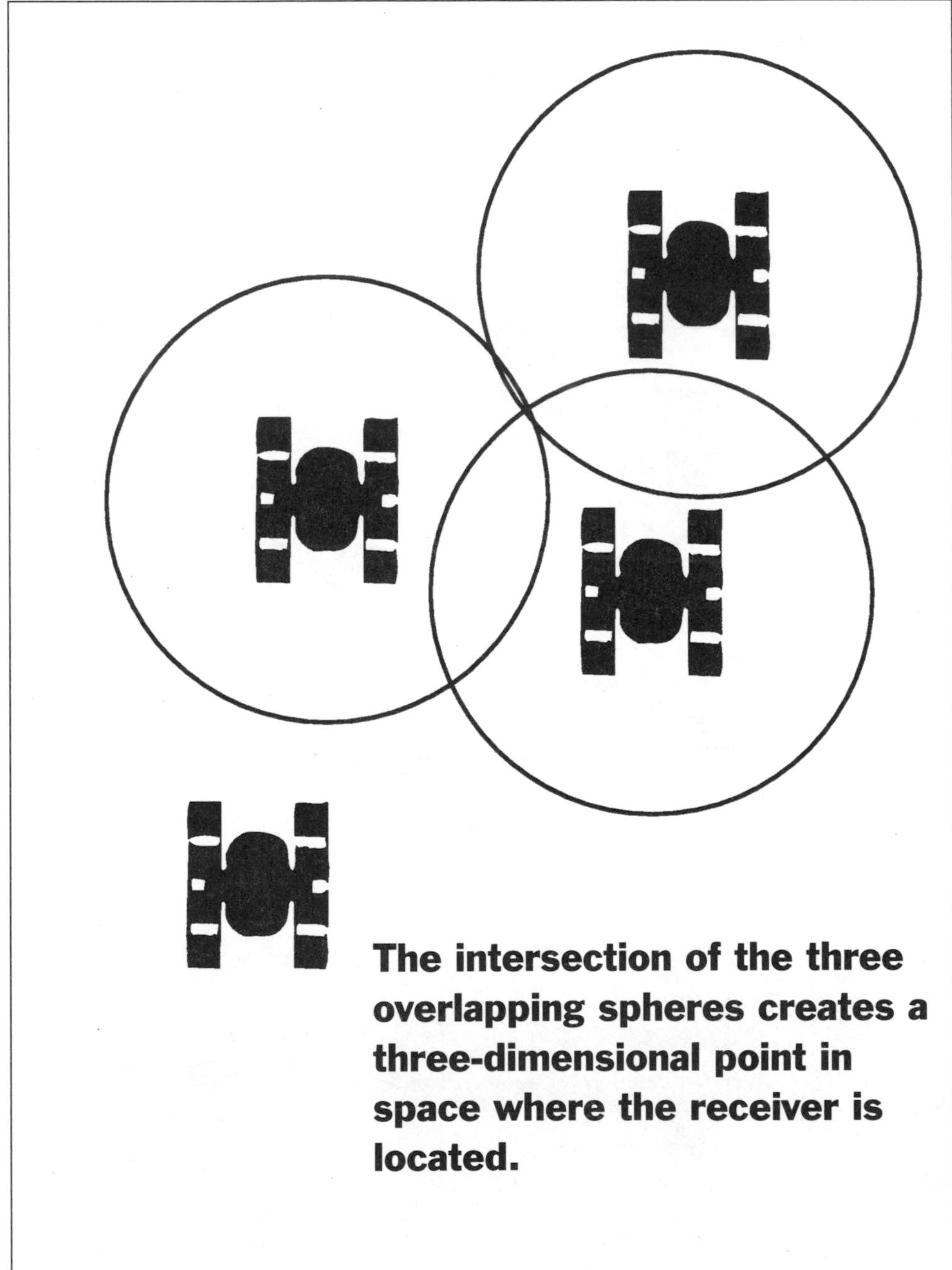
The intersection of the three overlapping spheres creates a three-dimensional point in space where the receiver is located.

Fig. 21.9

However, this is not accurate enough for Category I, II, and III precision approaches and autolandings.

DIFFERENTIAL GPS

In order to make the GPS more accurate, differential corrections can improve navigational accuracy from 328 feet to 32 feet. Using a differential GPS (DGPS) system, a special antenna is located at a fixed point within an area of interest (in the vicinity of an airport or airports) (figure 21.10). Signals are received and observed in real time and compared with the signals expected to be observed at that precise point. The differences between the observed signals and predicted signals are transmitted to users as differential corrections. These corrections are used to increase the precision of the GPS receiver. The differential facility is able to serve an area of a few miles to several hundred miles depending upon intended use and facility location. It can be possible to implement DGPS worldwide. This type of accuracy is what is needed for precision approaches and autolandings.

USES

The FAA has determined that GPS accuracy of 328.1 feet is suitable for all current civil aviation accuracy requirements with the exception of precision approaches and landings (Category I, II, III, and autolandings). GPS is currently in an implementation process.

The FAA is implementing GPS technology into civil aviation using a three-phase process. The first phase began on June 9, 1993, with the introduction of GPS overlay non-precision approaches. Phase I allowed the primary use of GPS for IFR operations providing that traditional navigation equipment monitored the GPS equipment. This monitor requirement was also required when GPS non-precision approaches were conducted. GPS overlay approaches were authorized by the F.A.A. for non-precision approaches consisting of VOR, VOR/DME, NDB, and other special non-precision instrument approaches. Essentially, Phase I allowed GPS operations while being monitored by traditional navigation facilities as a backup.

Phase II began on February 17, 1994, which allowed GPS to be the sole means of navigation as long as the GPS equipment meets FAA requirements. This phase dropped the monitoring requirement using traditional navigation facilities. These first two phases are in effect transparent to the air traffic control system. There are no operational changes required by the air traffic controller to incorporate either of the first two phases.

The third phase began when the GPS was declared fully operational. This took place in 1995. This phase will allow full and sole use on GPS equipment that meets FAA requirements. The approach procedure names will be changed to

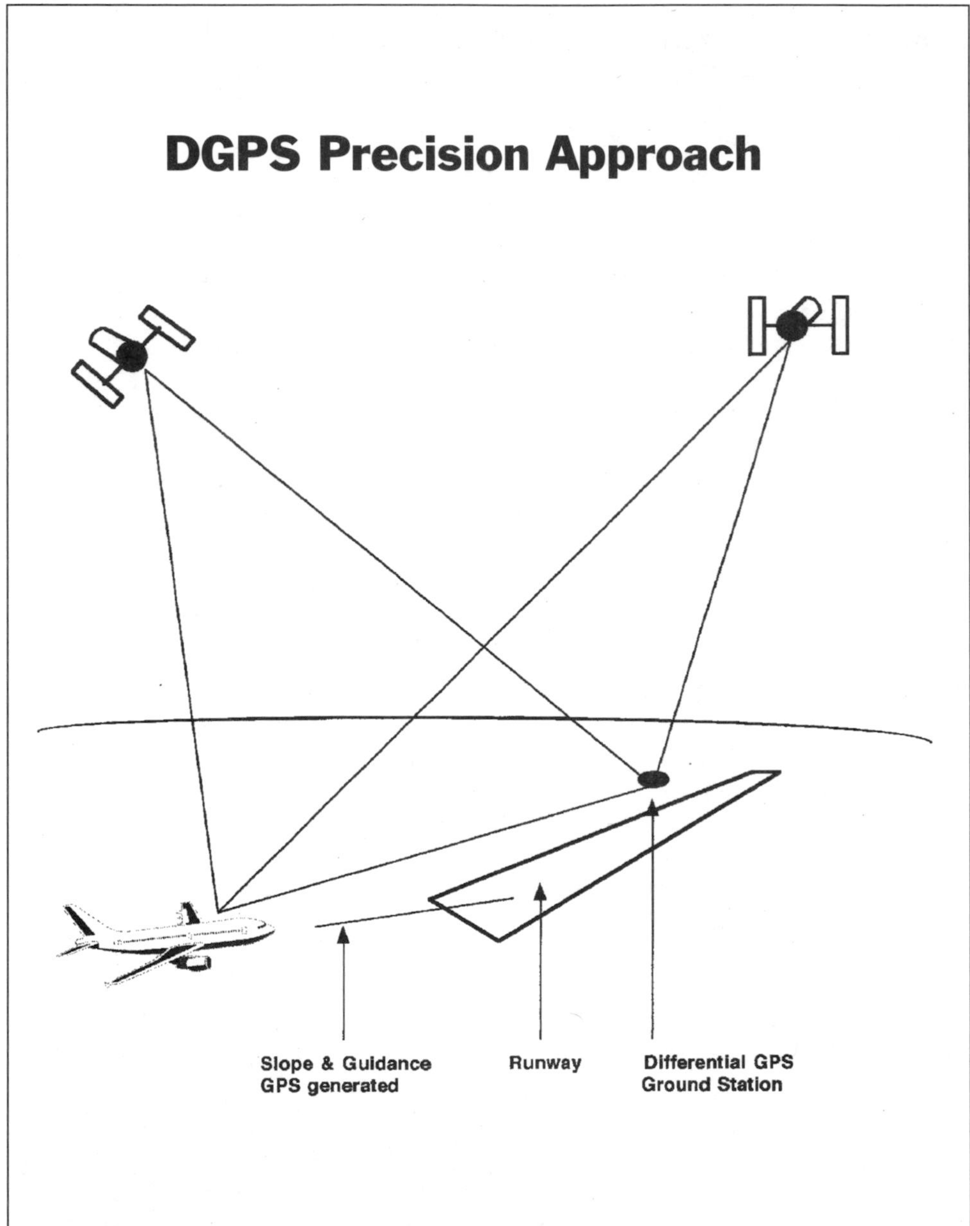
DGPS Precision Approach
Slope & Guidance
GPS generated
Runway
Differential GPS
Ground Station

Fig. 21.10

incorporate the specific procedure. For example, "GPS RWY 18" will be the phraseology for the clearance and for the name of the approach. This phase will delete the requirement for the ground navaid that supports the approach to be operational. This phase will begin specific phraseology for pilots and controllers for GPS approaches and clearances.

The GPS technology will revolutionize navigation as we know it today. Eventually all ground based navaids will become obsolete. GPS receivers will be in the cockpits of virtually all airplanes. The slow dismantling of ground based navaids will occur over time. Until that time, navigation will continue to be a mix of the different technologies that are available. GPS navigation is viable because it provides a low cost (relatively), very accurate, worldwide system of navigation. It is only a matter of time before GPS is *the* navigation system of the future.

22 Navigation in the glass cockpit

COMPUTERS HAVE INFLUENCED ALL AREAS OF OUR LIVES, AND AIR navigation is no exception. Today in larger airplanes and tomorrow in light airplanes, mechanical flight instruments will be replaced by computer screens. The screens are smaller versions of the monitors like those of today's personal computers. The screens are small television pictures made of glass, and that has coined the phrase "Glass Cockpit."

Navigation in the glass cockpit has similarities to conventional airplanes, and at the same time has some major differences. Navigation in the glass cockpit requires a basic understanding of the airplane systems that are providing the pilot with the data. It is also important to understand that navigation in the glass cockpit is not only horizontal or lateral, but vertical as well. The primary purpose of the glass cockpit is to reduce the pilot workload and increase fuel efficiency by integrating a variety of airplane systems. These systems include the flight management computer (FMC), the control display unit (CDU), the autothrottle (A/T), the autopilot flight director system (AFDS), and the inertial reference system (IRS). Integration of these systems is what is known as the flight management system (FMS) (figure 22.1).

FLIGHT MANAGEMENT COMPUTER

The flight management computer is a data base that contains information relating to the

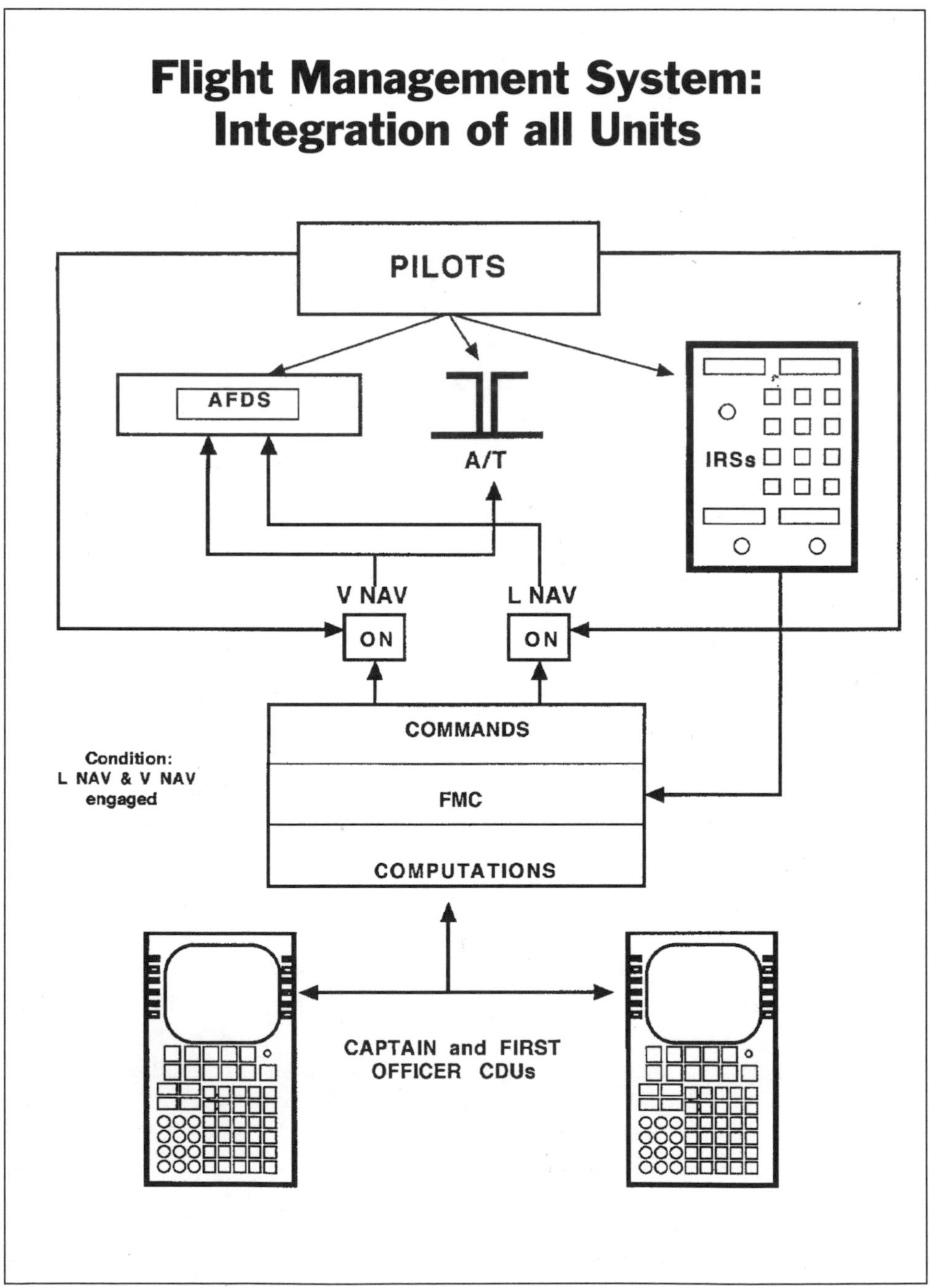
Flight Management System:
Integration of all Units
PILOTS
AFDS
A/T
IRSs
V NAV
L NAV
ON
ON
COMMANDS
Condition:
L NAV & V NAV
engaged
FMC
COMPUTATIONS
CAPTAIN and FIRST
OFFICER CDUs

Fig. 22.1

performance and navigation of the airplane. The FMC is accessed by the pilot through the control display unit. By using the CDU, the pilot is able to communicate with the FMC in a compatible language to find the information stored in the data bases, and then use the information in a manner that can accomplish the task at hand. The FMC and CDU combined provide the pilot with a flight management tool which performs navigational and performance computations. This is accomplished by accessing the two data bases stored in the FMC, the performance data base and the navigational data base (figure 22.2).

The purpose of the performance data base is to reduce the need for the pilot to refer to the airplane performance manuals. This data base also provides the autothrottles the information for thrust commands as selected by the FMC or by the pilot. This data base is also used by the FMC to provide predictions along the entire route of flight. An example of a prediction would be information about when to step climb to a higher altitude for increased fuel efficiency.

The performance data base contains information relating to the specific airplane and its operating characteristics. The performance data base contains the average model of the airplane and its engines. This data is compiled from the design process and flight certification tests, and has been shown to be accurate in the actual operation of the airplane. The average data base contains information pertaining to climb and cruise performance, maximum and optimum altitudes, certain drag characteristics, and maximum, minimum, and economy airspeeds for different operating conditions. Also contained in the data base is a buffet limit envelope, certificated operating limitations, and altitude and speed of single engine capability. This information is used by the pilot to make decisions about the operation of the airplane.

The performance data base, as well as the navigational data base, is accessed through an input/output process, thereby allowing the pilot to make entries and receive information based on that particular input. Once the gross weight of the airplane is known by the FMC, an economical climb speed is computed, and an economical altitude is presented along with an appropriate cruise speed. A descent airspeed is calculated based on weight at the top of descent and the winds aloft. The vertical outputs from the FMC are normally referenced to the best economy profile for climb, cruise, descent, and holding patterns. Computations of optimum airspeeds for the economy profile are based on a cost index, which is the ratio of operating costs compared to the cost of fuel. Changes in the cost index change the computed optimum airspeeds by the FMC. However, at any time, any speed profile can be selected by the pilot other than economy, by manual selection of the pilot.

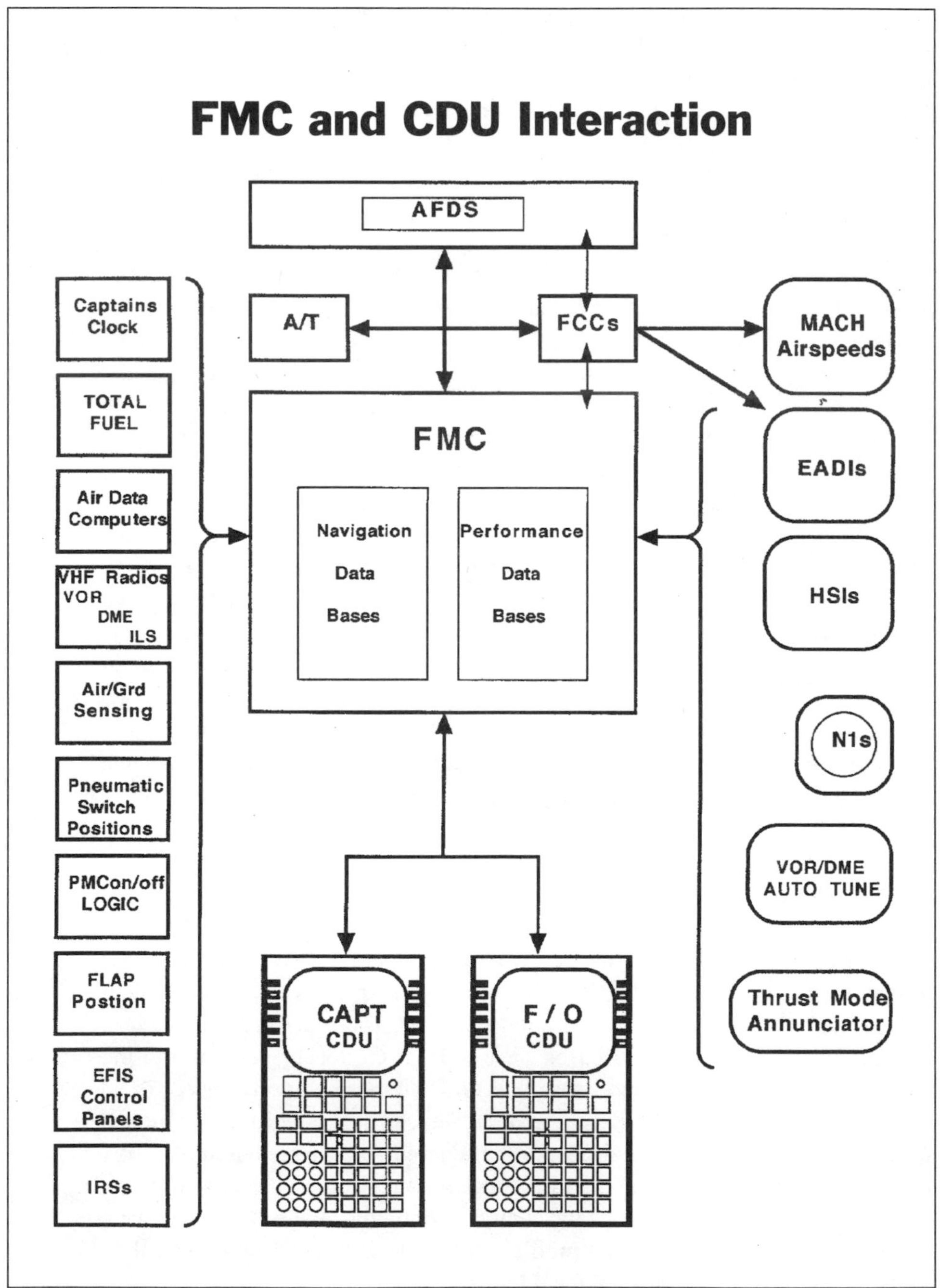
FMC and CDU Interaction
AFDS
A/T
FCCs
MACH Airspeeds
Captains Clock
TOTAL FUEL
Air Data Computers
VHF Radios VOR DME ILS
Air/Grd Sensing
Pneumatic Switch Positions
PMCon/off LOGIC
FLAP Postion
EFIS Control Panels
IRSs
FMC
Navigation Data Bases
Performance Data Bases
EADIs
HSIs
N1s
VOR/DME AUTO TUNE
Thrust Mode Annunciator
CAPT CDU
F / O CDU

Fig. 22.2

The navigation data base in the FMC is stored in two parts. One part of the data base is active data that is effective until a specific date of expiration. The other is a set of data that is to be used for the next period of effectivity. The navigational data base is kept up to date by maintenance personnel on a revision cycle that normally coincides with the date of flight manual and chart revisions. Through this method, the data in the FMC is current with the pilot's charts and maps.

The FMC navigational data base contains information for all VOR and DME stations in the United States, as well as selected ILS facilities, airports, and runways. Additional information is also stored in the data base, such as routes, airways, standard instrument departures (SIDs), standard terminal arrival routes (STARs), and instrument approach procedures. The data base can be applicable to any location in the world depending upon the software installed into the data base. The data base is tailored to the specific customer and their areas of operation. The data presented includes most of the information normally obtained from navigational charts. This information can be displayed on the CDU or the electronic horizontal situation indicator (EHSI), and thus eliminate the majority of chart reading that is done in conventional navigation airplanes. (This is not to say charts are to be abandoned in the glass cockpit. However, they are used in a verification process instead of initial course selection.)

There is a wealth of information located in the navigation data base for the pilot. When an airport is entered, the latitude and longitude, the station elevation, and magnetic variation of the field are displayed. When a runway is entered, the latitude and longitude, runway elevation, runway length, and ILS frequency are displayed. When entering a VOR, the frequency, the latitude and longitude, elevation, class of VOR, and magnetic variation are all presented to the pilot. The same information is available for non-directional beacons. When SIDs and STARs are used, courses, airspeeds and crossing altitudes, and all of the fixes of the transition are presented. When using an instrument approach, the inbound course, fixes, and altitudes are made readily available to the pilot.

The FMC and CDU provide the pilot with a tool that performs navigation and performance computations. These computations are both vertical and horizontal. Computations that relate to horizontal or lateral navigation include estimated time of arrival, distance to go, and courses to be flown in accordance with the route or airway. Route segments include published airways, constant headings, or great circle tracks between selected waypoints. Vertical navigation computations include fuel burn data for each phase of flight, when to step climb to a higher altitude, and airspeeds and altitudes to be flown. All of these items are determined by the FMC and displayed on the CDU.

CONTROL DISPLAY UNIT

The control display unit (figure 22.3) is the tool that lets the pilot interact with the FMC. Through the CDU the pilot is able to input to the FMC what actions are required. The FMC can then perform the task requested by the pilot, or inform the pilot via a message why the task cannot be completed. If the request is accepted by the FMC, the task is accomplished and no further communication is required. The CDU data is displayed to the pilot in the form of pages, which relate to the applicable phase of flight. These pages can be selected by simply pushing the button that corresponds to the phase of flight that the pilot is concerned with. The phase of flight selected need not be the one that is currently being flown. For example, during the climb, the pilot can select the cruise page and enter and execute a different cruise airspeed that will become active when the airplane reaches the cruise altitude.

The CDU is used throughout the flight from preflight to parking. During the preflight, the CDU is used to inform the FMC of the route to be flown, altitude selected, requested airspeed if desired by the pilot, gross weight, forecast winds, and latitude and longitude from the inertial reference system (IRS). These entries are the starting point for the FMC and the IRS. After takeoff, the FMC continuously updates the progress of the flight, with all data available through the CDU. Any modification to the progress of the flight can be entered through the CDU.

INERTIAL REFERENCE SYSTEM

The FMC is certified to be accurate within two nautical miles per hour when it is in range of accurate VOR/DME facilities. The FMC uses a combination of sources to determine the most accurate position. The position is established by using radio inputs and inputs from the inertial reference system (IRS). The FMC automatically crosschecks radio position, and updates the facilities every two minutes with a stronger signal. The FMC position can be derived from information from the IRS only. However, when in range of accurate radio inputs, the position is continually refined. The radio inputs can be from VOR/DME, ILS/LOC, or DME/DME. The IRS position is then checked against the radio position to determine the best position. Using this data the FMC determines the best position every five seconds.

The IRS provides the FMC with many sources of information that are essential to navigation. This system provides present position in terms of latitude and longitude, attitude, true and magnetic heading, groundspeed, track, and wind data to all systems that require inertial data. To provide this information, the IRS uses a system of laser gyros that use laser beam frequency shifts to measure movement of a particular axis, and accelerometers that detect acceleration.

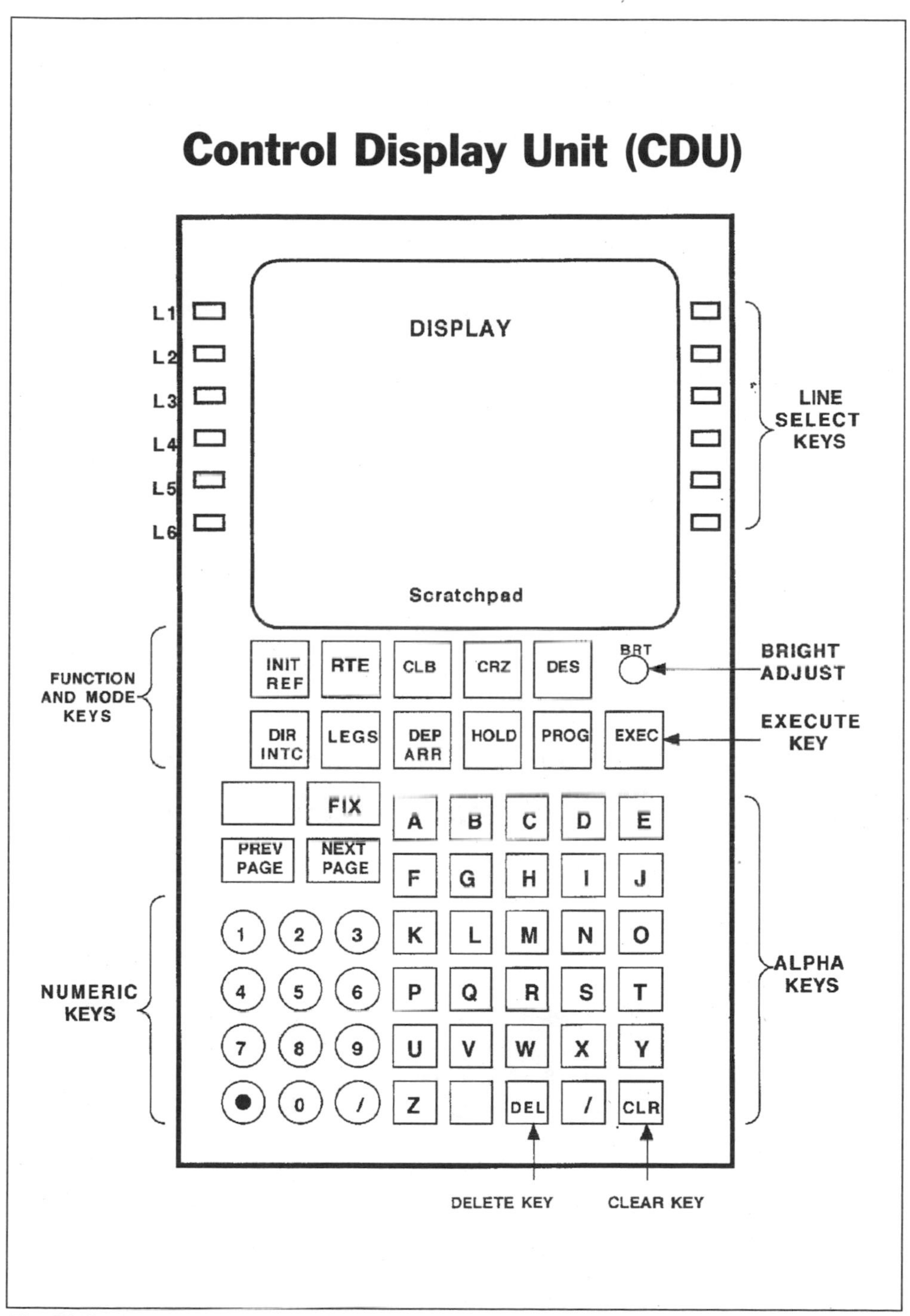
Control Display Unit (CDU)
DISPLAY
Scratchpad
L1
L2
L3
L4
L5
L6
LINE SELECT KEYS
FUNCTION AND MODE KEYS
INIT REF
RTE
CLB
CRZ
DES
BRT
BRIGHT ADJUST
DIR INTC
LEGS
DEP ARR
HOLD
PROG
EXEC
EXECUTE KEY
FIX
PREV PAGE
NEXT PAGE
A B C D E
F G H I J
K L M N O
P Q R S T
U V W X Y
Z DEL / CLR
1 2 3
4 5 6
7 8 9
0 /
NUMERIC KEYS
ALPHA KEYS
DELETE KEY
CLEAR KEY

Fig. 22.3

Inertial navigation can be defined as being able to determine the present location *without any assistance from outside references.* Inertial Navigation uses no radios at all! Each IRS contains three gyros and accelerometers. This is because the airplane can move in a three-dimensional plane. Each one of the gyros operates in conjunction with one particular axis. These gyros are referred to as ring laser gyros.

The ring laser gyro operates using two laser beams that travel in opposite directions (figure 22.4). These laser beams travel in a contained area in opposite directions toward the same point where a measurement is taken. Anytime there is rotation of the axis, one of the beams of light will shorten and the other will lengthen. The difference in the two beams' paths is measured by a detector from a designated measurement point and sent to a microprocessor (computer). The difference in frequency between the two beams is a measure of rotational rate about the axis. The microprocessor then takes the inputs received and applies the earth's rotational acceleration, the earth's spherical shape, and the effects of gravity to determine the system outputs.

The IRS uses three other inputs that assist in providing all of the outputs that are used by various airplane systems. These are initial position, barometric altitude, and true airspeed. The initial position is used by the IRS during the alignment process to become the navigation starting point. Although the IRS stores its last position in memory prior to shutdown, the position entered by the pilot serves as a verification and reference point for the alignment process. Barometric altitude stabilizes vertical navigation, which stabilizes vertical outputs to the system. The TAS allows the system to be able to determine wind direction and speed. The IRUs have been designed to operate with AC power while using DC power as a backup.

AUTOPILOT AND FLIGHT DIRECTOR SYSTEM

The autopilot and flight director system are integrated into the overall system. When the autopilot is engaged, it follows commands from the FMC. These commands include navigation and performance commands. The pilot selects a navigational command, enters it through the CDU to the FMC, the autopilot will follow those commands as long as the lateral (horizontal) navigation function of the autopilot is engaged. If the pilot changes a performance command, the autopilot will accomplish the required command as long as the vertical navigation function of the autopilot is engaged.

The flight director is presented to the pilot on the electronic attitude direction indicator (EADI). The flight director presents pitch and roll commands in the form of bars on the EADI. These commands are from the FMC for either the autopilot to follow, or for the pilot to follow when manually flying the aircraft.

Ring Laser Gyro

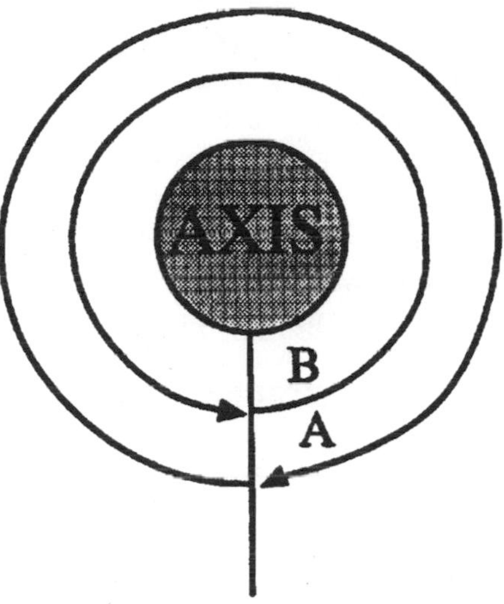

A & B are light beams that travel in different directions.

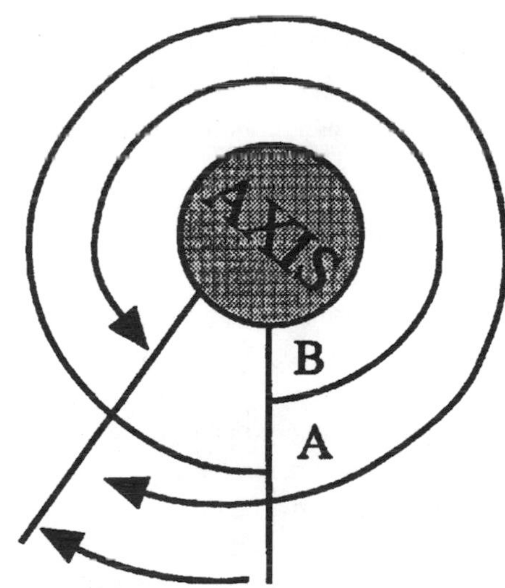

AXIS MOVEMENT

When the axis moves, the beams of light become different lengths. The Photocell Detector measures the difference. This difference is sent to the microprocessor that determines the axis movement.

Fig. 22.4

The flight director follows both lateral navigation and vertical navigation commands as computed by the FMC and selected by the pilot.

AUTOTHROTTLES

The autothrottles, or thrust management system, plays an important part of the vertical navigation function. The autothrottles, when engaged, receive commands from the FMC. The autothrottles are available from takeoff through landing, and provide the proper thrust based on the FMC commands. These commands can be changed at any time by the pilot through the CDU. These commands include proper takeoff thrust, climb thrust, cruise thrust, and thrust necessary to make crossing restrictions during descent.

ELECTRONIC FLIGHT INSTRUMENT SYSTEM

The electronic flight instrument system (EFIS) is actually what is known as "glass" in the cockpit. The term glass refers to the cathode ray tubes (CRTs) that are now used in the latest generation of airplanes. The EFIS consists of CRTs that encompass the electronic attitude direction indicator, the electronic horizontal situation indicator (EHSI), and sometimes include all of the engine instruments on a separate CRT. The EADI and EHSI receive attitude, heading, and track information from the IRS, and flight and map information from the flight management system.

The EADI (figure 22.5) is displayed much in the same way as conventional ADIs. However, using the CRT screen, much more information can be provided to the pilot. The ADI is able to present conventional information to the pilot such as attitude deviations, flight director commands, localizer and glideslope deviations, and some even present an airspeed display. Other information, however, can also be presented. Information such as TAS and groundspeed, radio altitude, decision height, a pitch limit symbol to alert the pilot of maximum pitch, and windshear information can be provided for the pilot. A wealth of information can be presented over and above that of a conventional ADI.

The EHSI (figure 22.6) is a major departure from conventional HSIs. Although the EHSI has selectable modes for the pilot, the one most commonly used is known as the map mode. The map mode presents information against a moving background as the airplane is moving. The map background moves relative to the fixed airplane symbol. Information presented includes current route, magnetic heading, trend vectors, range to altitude, wind speed and direction, distance to next waypoint and estimated time of arrival, and selected navigational facilities and airports. The map range can be changed from 10 miles up to 320 miles, with selections in between. Weather radar is also displayed on the map. This depicts weather in relation to the map range selected, and shows weather in the proper location in conjunction with the moving map. The pilot has the option of select-

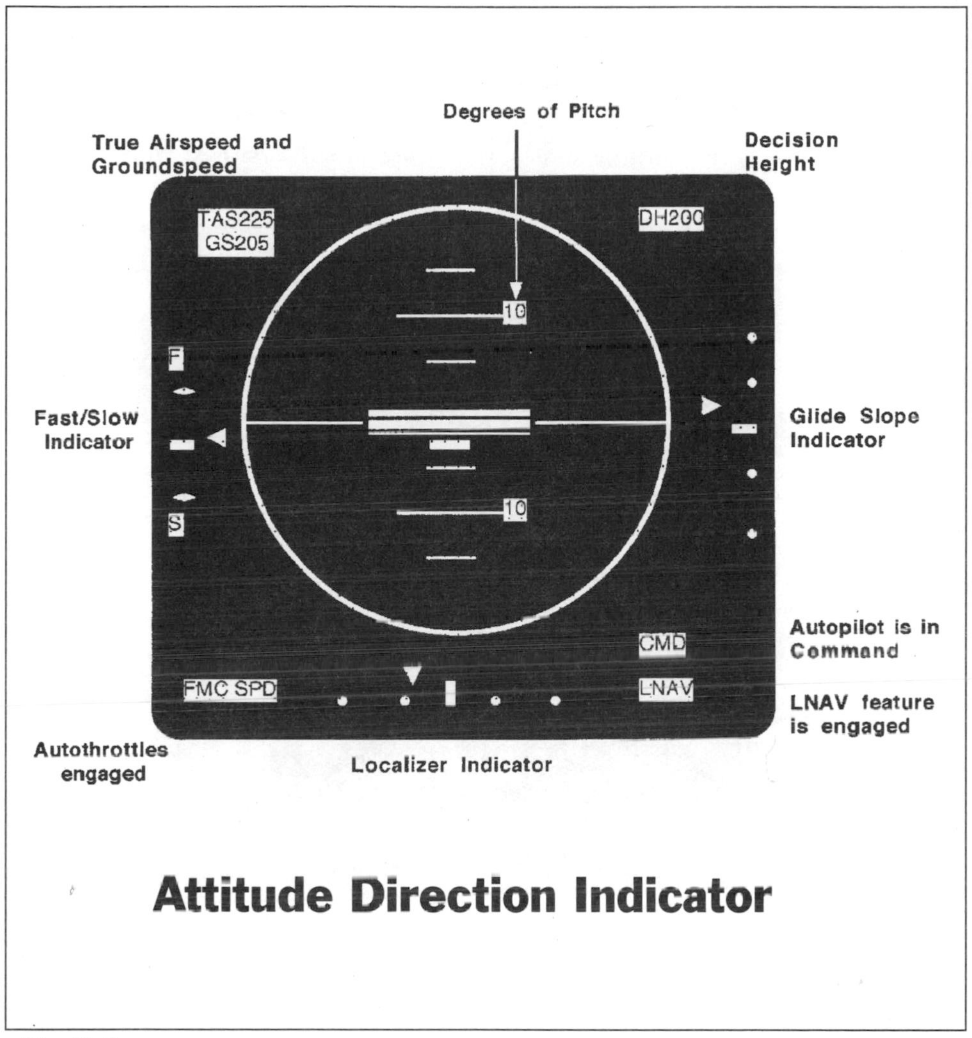

Fig. 22.5

ing conventional VOR or ILS modes for presentation instead of the moving map display.

The symbol generators are the instruments that provide the displays for the CRTs on the EADI and EHSI. The symbol generators receive inputs from the various avionic systems and process the information into displays for the appropriate CRT. Invalid or unreliable data results in an appropriate failure flag or removal of the affected display from the CRT.

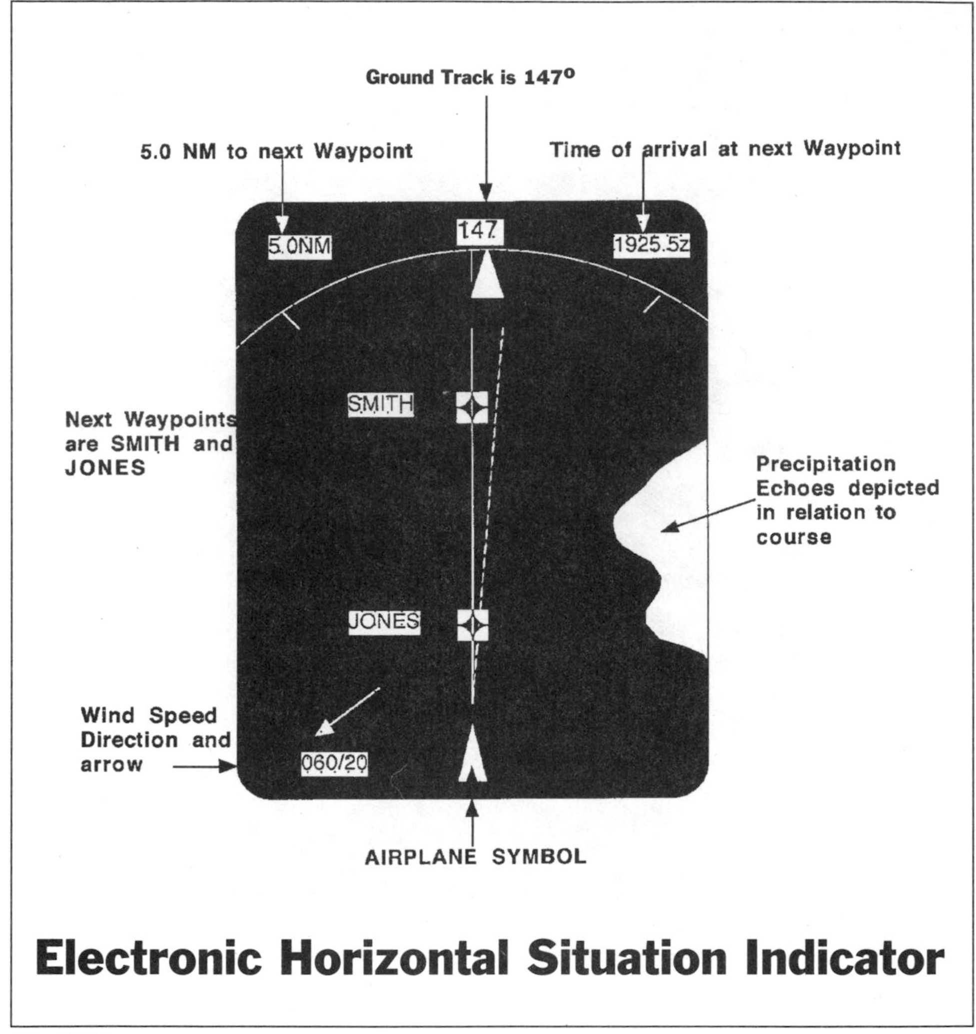

Fig. 22.6

NAVIGATION

The actual navigation performed in a glass cockpit airplane is quite simple when all the resources available are used. The pilot enters the route to be flown, desired altitude, and desired airspeed if economy is not chosen, and then the FMC computes the vertical and lateral computations. The route appears as a line on the EHSI, whether a direct routing or an airway. The line is then followed by the autopilot, or by the flight director if the pilot is manually flying, as long as the LNAV feature is engaged. The vertical navigation is accomplished by use of the autothrottles and pitch control, and engagement of the vertical navigation

(VNAV) function. The autopilot or flight director will command pitch to maintain the proper climb and descent airspeeds, along with proper thrust provided by the autothrottles. The cruise portion of the flight is also encompassed in the vertical navigation computation.

The route is entered through the CDU to the FMC. The FMC produces a line that corresponds with the entered route and displays the line on the EHSI through the symbol generator. The airplane navigates with reference to the line on the EHSI. Airplane airspeed is controlled by the autothrottles through FMC commands. When a fix requires a crossing altitude and airspeed, the throttles reduce to idle, the airplane pitches down and crosses the fix at the programmed altitude and airspeed. The pilot has become an operation overseer. The pilot simply has to make sure the FMC is programmed correctly and verify the airplane will accomplish what the FMC commands.

An example of how the FMS is now being integrated into the airspace system is an FMS transition (figure 22.7). These FMS transitions are for a particular airport and are runway specific. The track of the transition is the same track over the ground that the controller would use when vectoring airplanes for the final approach course. When an air traffic controller assigns an FMS transition, the pilot selects the runway and the specific transition through the CDU. The controller then gives the pilot a clearance direct to the initial fix or a vector to it. With the LNAV feature engaged the airplane will follow the course. The advantage of this type of procedure is that the pilot knows exactly the relationship of the airplane to the final approach course. The other advantage of these transitions is the reduction of talk on the radio. Radio congestion is becoming a major problem at busy airports.

The conventional method of flying airplanes is always available to the pilot in the glass cockpit. To regain complete manual control of the thrust, turn the autothrottles off. By placing the EHSI in the VOR mode, and manually selecting VORs and courses to be flown, the airplane can be flown by conventional methods. The features of the glass cockpit just make the overall navigation simpler.

CONCLUSION

Navigation is simply getting from here to there. In all navigational endeavors, the most important point is determining where "here" is. This is true whether pilotage, dead reckoning, VOR, LORAN-C, or GPS navigation is used. Once "here" is established, the process of getting to "there" can begin.

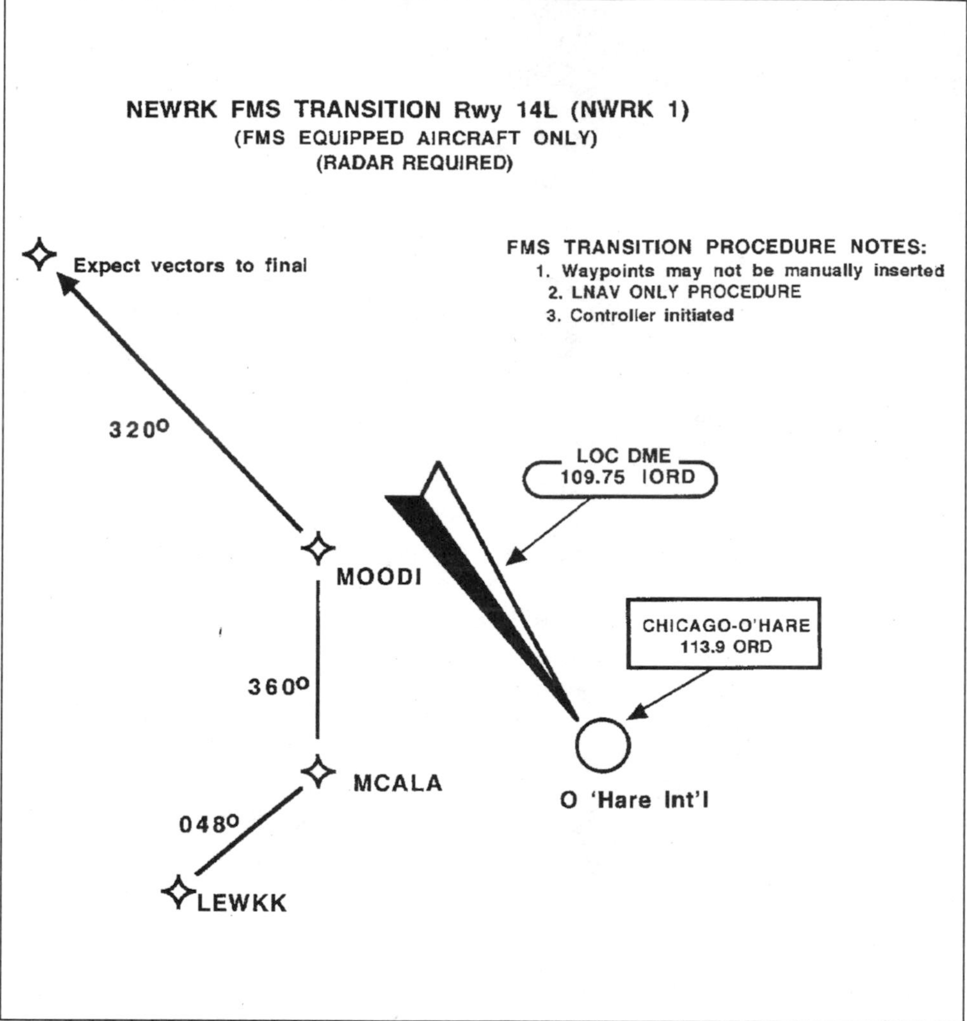
NEWRK FMS TRANSITION Rwy 14L (NWRK 1)
(FMS EQUIPPED AIRCRAFT ONLY)
(RADAR REQUIRED)
Expect vectors to final
FMS TRANSITION PROCEDURE NOTES:
1. Waypoints may not be manually inserted
2. LNAV ONLY PROCEDURE
3. Controller initiated
320°
LOC DME
109.75 IORD
MOODI
CHICAGO-O'HARE
113.9 ORD
360°
MCALA
O 'Hare Int'l
048°
LEWKK

Fig. 22.7

23
Airspace rules and collision avoidance

WHY IS A CHAPTER ON COLLISION AVOIDANCE INCLUDED IN A navigation book? Most pilots consider the study of navigation and the study of airspace rules as separate topics. But did you ever ask yourself what the real reason for all the airspace rules actually is? Simply stated, the only reason we have different types of airspace is to help keep airplanes from running into each other!

Nearly all collisions take place in VFR conditions. Regardless of whether a pilot is on an IFR flight plan, a VFR flight plan, or no flight plan, the pilot must look out the window when in Visual Meteorological Conditions (VMC). The regulations place the responsibility squarely on the pilot: "Vigilance shall be maintained by each person operating an aircraft so as to see and avoid other aircraft." The pilot should be ever vigilant against flight path intrusion. This idea is tough to bring to reality when the pilot's job is also cluttered by charts, radio transmissions, instruments, and passengers. Even if all the pilot did was look outside, there are large chunks of air that are blocked from view by the aircraft itself. But the burden must lie somewhere. It lies not with the controllers but with the pilots.

SEE AND AVOID

You might find it a little scary to learn that the controllers are under no obligation to prevent aircraft targets on the radar screen from intersecting. If the targets are on an IFR flight plan, the chances are better that the pilot will get a traffic advisory, but traffic advisories are given on a time-share basis. When a controller is working with traffic on a VFR day, and out of the corner of his eye he sees two targets nearing each other on another part of the screen, he just has to think to himself, "They will probably miss each other, but it is their problem, not mine." Do not get the idea that controllers are callous to this situation: They are not. But controllers can only do so much. What is done is prioritized. VFR traffic advisories are on the priority list, but they are not at the top of the list.

So, *see and avoid* is the way of flying. Simply put: The pilot must first see other aircraft and then do whatever is necessary not to hit other aircraft. No matter how high-tech the system has become, this simple fact remains: If a pilot wants to avoid a midair collision, he must not run over anybody or be run over. Nobody else can be counted upon to help the pilot accomplish this job, and nobody else will ultimately be responsible for the job. This applies to small airplanes, airliners, military aircraft, everybody.

The challenge now becomes seeing the traffic and avoiding situations where traffic cannot be seen. A pilot's eyesight is extremely valuable to him. But good eyesight alone will not guarantee vigilance. A pilot may have 20/20 uncorrected vision but not know how to use his eyes properly. Airspace rules are written to give the pilot the best chance to see traffic, but the pilot must finish the job. The pilot must spot the danger.

"Vigilance shall be maintained by each person operating an aircraft so as to see and avoid other aircraft."

During flight, the pilot must divide his attention between what is outside and what is inside. Use a personal "time-share" arrangement with your eyes. Do not spend long periods of time on any one item. Follow a pattern similar to this:

1. Look outside to the left and rear.
2. Look at the flight instruments.
3. Look outside down the left wing.
4. Look at the flight instruments, or make a notation on the nav log.
5. Look outside over the instrument panel out the front.
6. Look at the engine instrument, or check a chart position.
7. Look outside down the right wing.
8. Look at the flight instrument, or change a frequency.
9. Look outside to the right rear.
10. Start over again with number one and repeat.

The more proficient the pilot is in working the radios, using charts, using the Navigation Record, and identifying problems on the instrument panel, the more time is left over for looking outside. When the pilot looks outside, he or she should not "sweep" the sky, but rather look in one direction long enough for the eyes to focus. The eyes do not focus while they are in motion. If a pilot looks up momentarily from the problem of navigation and swings his head from side to side in the name of collision avoidance, nothing has been accomplished. Even 20/20 eyesight will not help if the eyes are not allowed to work. When the eyes are directed outside, they will focus on the first thing in the field of vision. If the first thing is a bug on the windshield, the eyes will focus on the bug and temporarily miss what is on the other side of the bug, namely another airplane.

The proper use of the eyes is just part of the story. The rest is airspace rules. The goal of the airspace rules is to give the pilot the best opportunity to see and avoid air traffic. Here is where a study of airspace, aeronautical charts, and regulations all entwine. Most well-prepared pilots can easily recite the basic VFR minimums but have a hard time putting them into practice, much less understanding their real purpose.

THE CONTROLLED VERSUS UNCONTROLLED AIRSPACE MISUNDERSTANDING

Essentially, controlled airspace exists wherever IFR traffic can be present. There are some exceptions to this statement, which will be covered later.

Whenever both VFR and IFR traffic coexist in the same airspace, there is the possibility of conflict. The conflict may be resolved if pilots see each other in enough time to avoid a collision. The basic distance from clouds in "controlled" airspace above 1,200 feet AGL and below 10,000 feet MSL is "500 feet below; 1,000 feet above; 2,000 feet horizontal" (figure 23.1). Many pilots believe that these distances are provided so that VFR pilots do not inadvertently find themselves inside clouds. They have been told horror stories about VFR pilots losing aircraft control within seconds after entering clouds, and they feel these distance rules are in place so they will stay away from the danger. The separation limits are thought to be merely exaggerated safety factors that can be ignored by the competent pilot. The danger of spatial disorientation of VFR pilots in IFR conditions is certainly real, but that is not why the cloud distance rule exists.

Refer to figure 23.2. In situation 1 of this figure, two airplanes are approaching head-on. The regulations say, "When aircraft are approaching head-on or nearly so, each pilot shall alter course to the right." This works great if the pilots see each other.

Assume that the distance between aircraft A and aircraft B of situation 1 in figure 23.2 is three statute miles. Three statute miles is the basic VFR visibility minimum in controlled airspace below 10,000 feet MSL. The FAA is telling the pilot with this rule that if two airplanes are within one and one half miles of a collision (three miles apart but converging at the same speed), there is plenty of time to see and avoid. These are very bare safety minimums! The closure rate of two Cessnas would be approximately 200 knots. This means that collision will occur in 45 seconds. A Piper with a speed of 100 knots and a Lear jet with a speed of 250 knots (the speed limit below 10,000 feet MSL) would have a closure rate of 350 knots. If the pilots in this situation were flying with only three miles visibility, the time to recognize the problem and take evasive action would be just 26 seconds. The 26 seconds would start when the airplanes were first visible to each other, at three miles apart. But the pilots probably will not see the other's plane exactly when it first comes into view. The time interval between when the other plane is visible and when it is actually seen will reduce the reaction time. If either pilot is scanning as quickly as once around every 15 seconds, then the time interval for evasive maneuvers could only be approximately 11 seconds. If the pilots were scanning only as often as 30 seconds, they could collide without ever seeing each other. Most pilot scans take longer than 30 seconds. The FAA thinks this situation is O.K. Pilots get lulled to complacency by thinking that everything that is legal is also safe. This is certainly not the case.

Now look at situation 2 of figure 23.2. These planes are not head-on, but they are on a collision course with each other. Their actual distance apart is 2,000

VFR visibility and distance from clouds				
Altitude	Uncontrolled airspace		Controlled airspace	
	Flight visibility	**Distance from clouds**	****Flight visibility**	****Distance from clouds**
1200′ or less above the surface, regardless of MSL altitude	*1 statute mile	Clear of clouds	3 statute miles	500′ below 1000′ above 2000′ horizontal
More than 1200′ above the surface, but less than 10,000 MSL	1 statute mile	500′ below 1000′ above 2000′ horizontal	3 statute miles	500′ below 1000′ above 2000′ horizontal
More than 1200′ above the surface and at or above 10,000 MSL	5 statute miles	1000′ below 1000′ above 1 statute mile horizontal	5 statute miles	1000′ below 1000′ above 1 statute mile horizontal
	*Helicopters may operate with less than 1 mile		**Control zone ceiling not less than 1000′	

Fig. 23.1

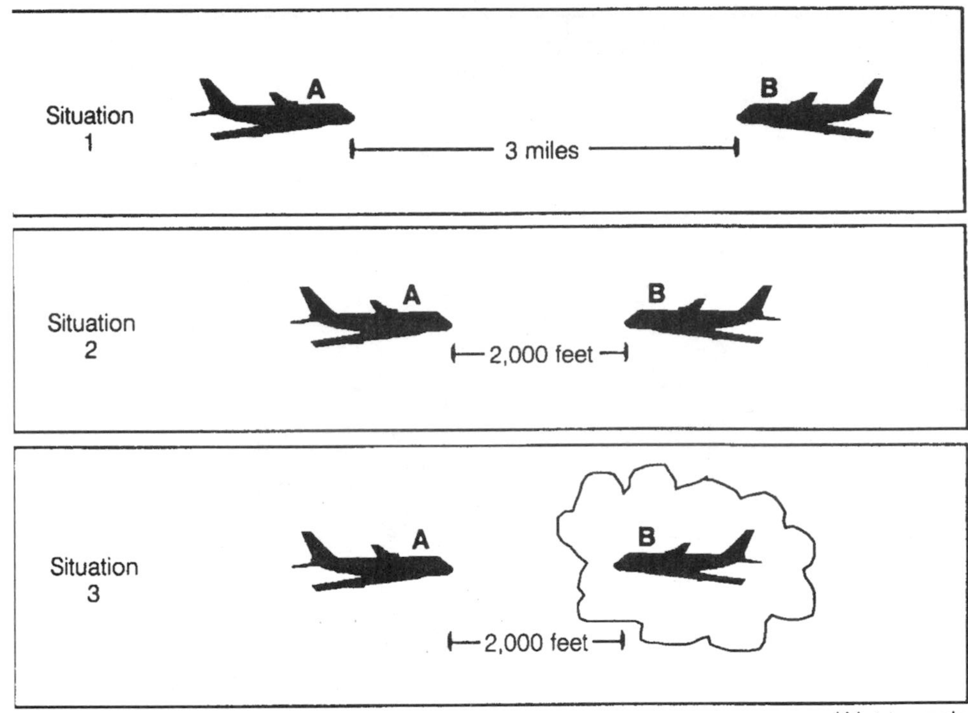

Fig. 23.2 *Not to scale

feet or less than one half statute mile. If at this distance the pilots first saw each other there would be virtually no time to miss each other. It was discovered by autopsy that the pilot of the Piper in the midair collision over Cerritos, California, experienced a heart attack prior to impact. The conclusion is that the man looked up to see a windshield full of DC-9 seconds prior to collision, but with no time to avoid the other plane, had a heart attack in fear. This is chilling, to say the least. Nobody would purposely place himself so close to another airplane that there is no time to avoid a collision. But in situation 2, an unavoidable collision is about to occur.

In situation 3 of figure 23.2, all factors are the same as in situation 2 except one. One of the airplanes is on an IFR flight plan at an assigned altitude. While on this clearance, the pilot will hold altitude and fly through whatever clouds stand in the way. The other pilot is on a VFR flight plan or no flight plan. Airplane B is inside a cloud and is doing what he is told. Airplane A has ventured to within 2,000 feet horizontally of the cloud. The airplanes are on a collision course but not necessarily head-on. Within one second, airplane B will pop out of the side of the cloud. Fifteen seconds after that, with a closure rate as slow as 150 knots, there will be impact. The closer that airplane A gets to the cloud, the less time will exist to get out of the way. A pilot who flies just clear of clouds will have absolutely no time to avoid traffic. Forget about the dangers of VFR pilots flying in the clouds. The real danger is that the clouds can be filled with moving metal!

Any time a VFR airplane flies in controlled airspace below 10,000 feet MSL these rules must be followed — and not because there is a risk of being caught. The question always comes up, "Who is going to know if I fly closer than 2,000 feet? Is there a police plane behind the cloud waiting to give me a ticket?" No, the FAA cannot enforce this one. But there is a death penalty!

If you fly above 10,000 feet MSL, you enter an area of no speed limits short of the speed of sound. In order to provide legal (but maybe not safe) time intervals at these altitudes, the VFR minimums expand to 5 miles visibility and 1,000 feet below; 1,000 feet above; 1 mile horizontal. One mile separation between a head-on Cessna with a speed of 100 knots and a Boeing 727 going 400 knots allows 6 seconds to first see then avoid. The FAA feels this is O.K. Here is where you must adopt personal minimums that are more stringent than FAA rules, if you feel that 6 seconds is not enough time.

A MIXTURE OF VFR AND IFR TRAFFIC

If you fly in uncontrolled airspace below 1,200 feet AGL (700 feet AGL in some places), then you are allowed to reduce the VFR minimums. In these locations, just one mile visibility is required and pilots can fly as close to clouds as they wish as long as they do not actually enter the clouds. The reason this is allowed is that there is less chance for IFR traffic to pop out of the clouds with little or no time to miss each other. The chances of collision are reduced but not completely eliminated. There is a loophole in the regulations that allows IFR flights in uncontrolled airspace without a clearance. FAR 91.173 says, "No person may operate an aircraft in controlled airspace under IFR unless they have filed an IFR flight plan and received an appropriate clearance." Nothing is said about a flight plan and clearance in uncontrolled airspace. This is a clear case of a situation where what is safe and what is legal are miles apart. While flying legal VFR in uncontrolled airspace with one mile visibility and clear of clouds, you could run into a legal IFR flight coming out of those clouds.

There is another, more common danger at airports with non-precision approaches such as VORs and NDBs. Very often the Minimum Descent Altitude (MDA) of a non-precision approach passes the IFR pilot from controlled to uncontrolled airspace. This allows for an IFR pilot to pop out of the MDA while at the same time a VFR pilot flying with one mile visibility and clear of clouds can be skirting the cloud bases in the traffic pattern. Both pilots are flying their airplanes in accordance with the FARs, but both are in jeopardy.

What decisions must be made about airspace when the visibility goes below three miles? The best decision for VFR pilots is to stay on the ground. There may be a legal way to fly, but this may start a bad decision chain. If flight is attempted, the pilot must at all times be aware of the boundary between controlled and uncontrolled airspace. On a Biennial Flight Review, I once asked a man to get his chart out and point out some uncontrolled airspace. He said he could not do that because he only had a Charlotte Sectional with him. I asked why that would be a problem and he said, "Because there is no uncontrolled airspace east of the Mississippi River." We had a long talk about airspace.

The truth is that uncontrolled airspace is just about everywhere, but especially down low. That makes sense because controlled airspace is used to protect the IFR pilot, and IFR pilots in the clouds cannot fly low because they cannot see obstructions while flying in Instrument Meteorological Conditions (IMC).

To determine where controlled airspace is, and is not, look at a sectional chart. The sectional chart has magenta colors surrounding many airports. These are referred to as transition areas (figure 23.3). Anywhere on the chart where the plane's position is surrounded by the reddish, magenta-colored shading, the controlled airspace begins at 700 feet above the ground and goes on up. Everywhere outside the magenta shading, the controlled airspace begins at 1,200 feet above the surface and travels on up.

Verify these questions using figure 23.3. (Answers are on page 202.)

1. An airplane is flying directly above the Carmi Airport. The airplane's altimeter is reading 1,100 feet. Is this airplane flying in controlled (Class E) airspace, or is the airplane flying in uncontrolled (Class G) airspace?

2. An airplane is flying directly above the McLeansboro Airport. The airplane's altimeter is reading 1,100 feet. Is this airplane flying in controlled (Class E) airspace, or is the airplane flying in uncontrolled (Class G) airspace?

3. An airplane is flying over the Carmi Airport while the visibility is only 2 miles due to haze conditions. The flight is taking place during daylight hours and the airplane's altimeter is reading 1,250 feet. Is this airplane operating legally for VFR?

4. An airplane is flying over the McLeansboro Airport while the visibility is only 2 miles due to haze conditions. The flight is taking place during daylight hours and the airplane's altimeter is reading 1,250 feet. Is this airplane operating legally for VFR?

The Floor of Class E Airspace as shown on a Sectional Chart

The airspace shown against the chart features (terrain, airports, highways, etc)

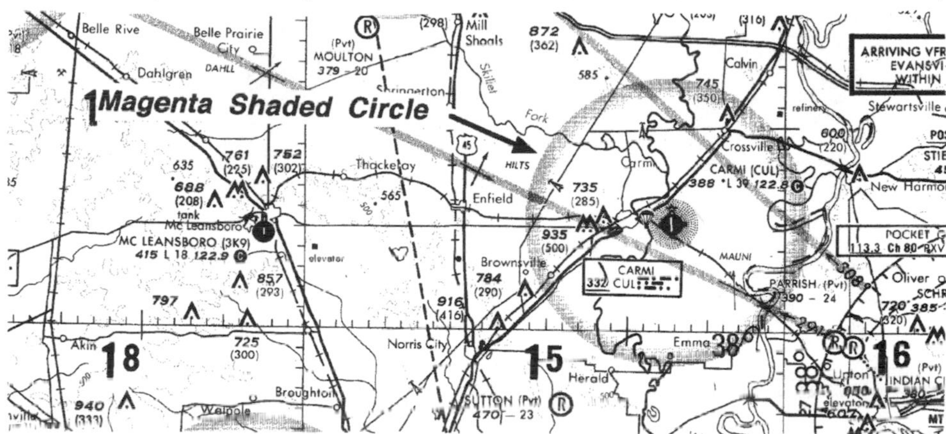

The same area of airspace shown here without the chart features underneath

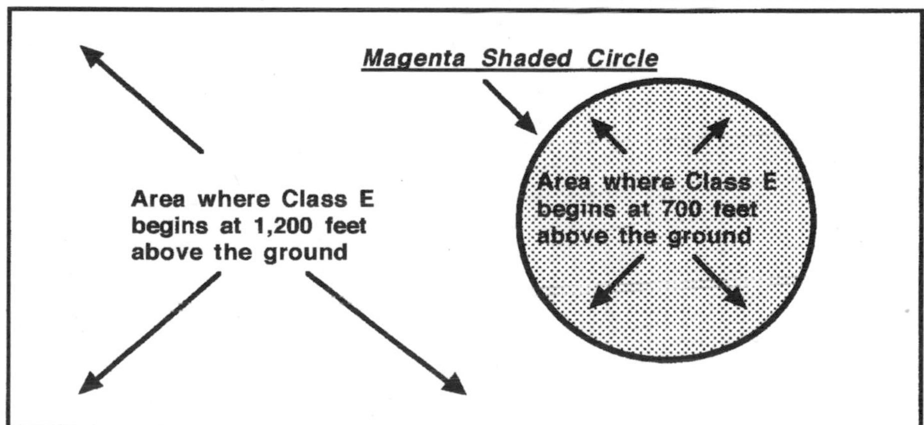

Fig. 23.3

24
The airspace boundaries and Class E airspace

WE KNOW THAT THE FLIGHT RULES WITHIN CONTROLLED AIRSPACE are different than the rules within uncontrolled airspace (remember figure 23.1). We also know that being in the wrong place at the wrong time can be deadly (remember figure 23.2). So how can a pilot know which rules to use when? How can a pilot determine when he is in uncontrolled or controlled airspace? Where is the boundary between the two?

Start by assuming that if you are standing on the ground you are in uncontrolled airspace and that controlled airspace is over your head somewhere. Except for designated areas the altitude where uncontrolled airspace stops and controlled airspace starts is 1,200 feet above the ground. The uncontrolled airspace has been designated as "G" airspace. The G is easy to remember because this type airspace is closest to the Ground. Above the 1,200-foot AGL boundary is controlled airspace and it has been designated as "E" airspace. While down low in G airspace the pilot can operate in the daytime with only one mile visibility and fly around just clear of the clouds. But, if the pilot climbs too high and passes above the boundary into E airspace, then three miles visibility is required plus the 500-, 1,000-, and 2,000-foot cloud separations. If the pilot accidentally gets too high and enters E airspace when there is only 1 mile visibility, then that pilot is in violation of the regulations, and worse he is in danger of not being able to see traffic before it is too late.

This boundary exists at 1,200 feet above the ground (AGL). But there is a problem. The altimeters in our airplanes indicate above mean sea level, not AGL. The altimeter alone cannot be used to determine where the G and E airspace boundary is located. What if the ground is 2,000 feet above sea level — where is the G/E airspace boundary? The boundary would still be 1,200 feet above the ground.

An airplane's altimeter would read 2,000 feet while on the ground in this location. If the boundary is 1,200 feet above the ground, then the boundary over this location would be 2,000 feet + 1,200 feet or 3,200 feet MSL. So at this spot 3,200 feet MSL is also 1,200 feet AGL!

The Earth's terrain is not flat. There are hills, valleys, mountains, cliffs, and plains. The G/E airspace boundary hovers 1,200 feet above all these terrain features. When a mountain juts up, the G/E airspace boundary juts up as well. When the Earth has a deep valley, the G/E airspace boundary dives into that valley, always staying 1,200 feet away from the surface. Can an airplane cross the G/E airspace boundary without climbing? Yes! An airplane can be flying straight and level in G airspace and while on the same indicated (MSL) altitude fly into E airspace. Say the airplane is holding an indicated altitude of 2,000 feet (figure 24.1). The ground's elevation below the airplane is 1,000 feet. There is a 1,000 foot separation between the ground and the airplane. The G/E airspace boundary is up higher at 1,200 feet above the ground. The airplane is therefore in Class G airspace. Then while the airplane holds the same 2,000 foot indicated altitude, the ground lowers into a valley. The ground's elevation under the airplane is now only 500 feet MSL. This means that the separation between the ground and the airplane is now 1,500 feet. The G/E boundary has lowered as the terrain lowered. The airplane is now in E airspace, even though its indicated altitude has not changed!

The G/E airspace boundary is a mirror image of the Earth's terrain features. The only difference is the 1,200 foot gap between the actual Earth and the mirror image of the Earth.

At various places, there are holes cut out in the mirror. The holes would mainly be above airports and have the magenta shading in them that would always remain 700 feet above the ground. As the pilot (in his imagination) looked out across the top of the mirror, there would appear to be many wells where controlled airspace (and all its rules of visibility and cloud separation) would be allowed to get a little closer to the surface. An uneven surface with holes all around, it might resemble a putting green. The bottom of the holes would not extend to the surface, but would rise and fall with the terrain and always be 700 feet above the terrain. So when the pilot flies into a position that is surrounded

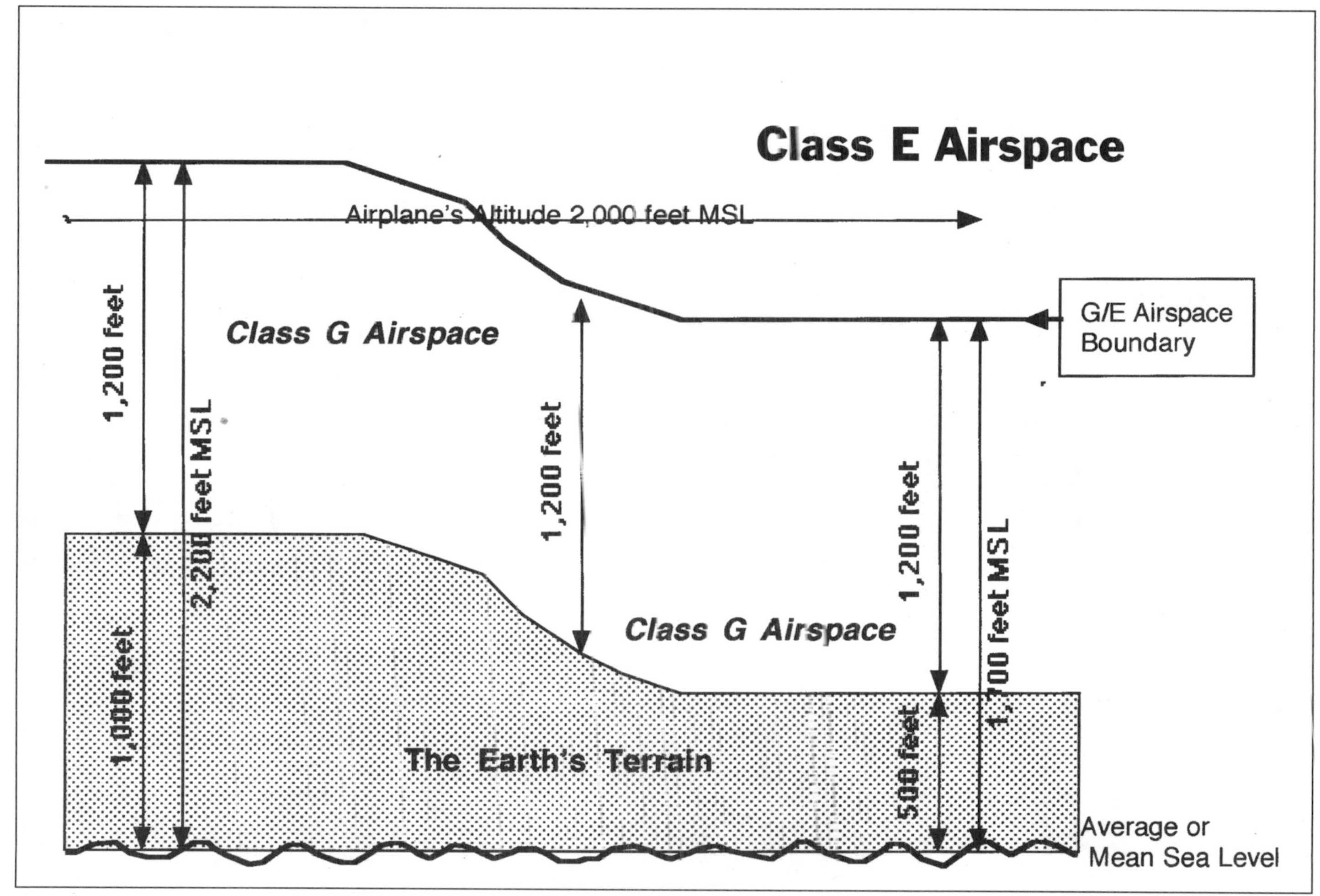

Class E Airspace
Airplane's Altitude 2,000 feet MSL
G/E Airspace Boundary
1,200 feet
2,200 feet MSL
Class G Airspace
1,200 feet
1,200 feet
1,700 feet MSL
Class G Airspace
1,000 feet
The Earth's Terrain
500 feet
Average or Mean Sea Level

Fig. 24.1

by the magenta shading on the chart the G/E airspace boundary drops down to 700 AGL.

A pilot on an IFR clearance, popping in and out of the clouds above the mirror, can safely descend into the magenta, passing in and out of clouds without fear of collision from VFR traffic. This gives the IFR pilot the best opportunity to get underneath the clouds and into the clear in order to avoid traffic and make a landing. This allows the IFR pilot to "transition" from IMC to VMC with reduced collision fears: hence the name Transition Area.

CLASS E TO THE SURFACE

In some places, the mirror with the magenta wells would have another hole cut out. This hole would lie inside the first hole. This hole would cut all the way to the ground and in doing so the controlled airspace (and all its rules of visibility and cloud separation) would drain down to the surface. This deeper hole that goes to the ground is where Class E Airspace touches the ground and there is no G airspace.

Not all airports that have Class E to the surface have precision approaches, but many do. The purpose of precision approaches is to allow an IFR pilot to use an Instrument Landing System (ILS) approach to the airport while in controlled airspace. An ILS approach can bring a pilot down to within 200 feet of the ground. If the airplane pops out with only 200 feet between the wheels and the ground, there is no room remaining for collision avoidance with a VFR pilot using the clear of clouds rule. So a Class E to the surface's first reason for being is to accommodate ILS approach traffic. The second reason is to protect IFR pilots on non-precision approaches from entering uncontrolled airspace prior to reaching the MDA and hopefully clear conditions. The transition from a non-shaded area to magenta shaded area to Class E to the surface (indicated on a chart by dashed magenta lines) offers the IFR pilot a staircase to fly down to a landing (figure 24.2).

Pilots not only use the sectional chart to find their way to the destination airport, the chart is also used to determine where the airspace rules change. Knowing what type of airspace the airplane is flying in is the pilot/navigator's responsibility. This knowledge keeps the flight legal, but most importantly it keeps the flight safe.

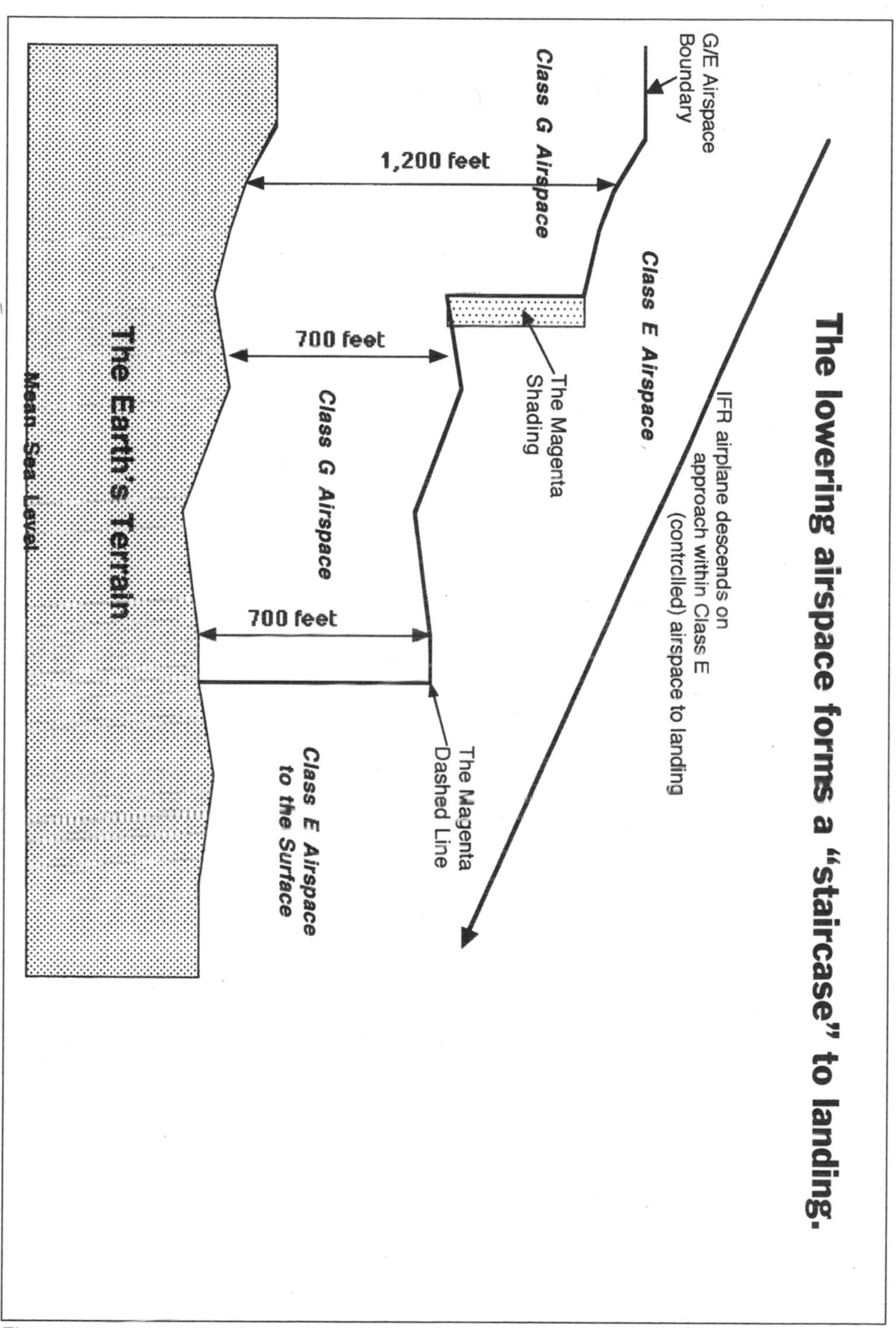

The lowering airspace forms a "staircase" to landing.
IFR airplane descends on approach within Class E (controlled) airspace to landing
G/E Airspace Boundary
Class G Airspace
1,200 feet
Class E Airspace
The Magenta Shading
700 feet
Class G Airspace
700 feet
The Magenta Dashed Line
Class E Airspace to the Surface
The Earth's Terrain
Mean Sea Level

Fig. 24.2

25 Special VFR and FSS communications

THE "CLASS E TO THE SURFACE" IS SAID TO BE IN EFFECT SOMETIMES and not in effect other times. This is a source of confusion. The Class E to the surface plays its role for traffic separation during times when the visibility goes below three miles and/or the cloud ceiling is less than 1,000 feet AGL. Then the reasons stated earlier for IFR approach protection are required. But a Class E to the surface does not go into effect until an FAA-qualified weather observer is in position at the airport within the Class E to the surface area. When the weather observer arrives, the base of controlled airspace actually moves. If the airport is under a transition area, the floor of controlled airspace is 700 AGL. When the weather observer arrives, the Class E to the surface goes into effect and controlled airspace lowers to the surface. When the observer goes home, the airspace jumps back up to 700 AGL. The weather observer makes the decision whether or not IFR conditions exist within the Class E to the surface area. If the observer determines that the visibility is less than three miles and the ceiling is less than 1,000 feet, he can declare the Class E to the surface to be IFR. If the Class E to the surface is IFR, and since basic VFR minimums require at least three miles visibility, the Class E to the surface becomes off limits to all but arriving and departing IFR traffic.

If the visibility is less than three miles, then the only place you may fly VFR is in uncontrolled airspace below 10,000 feet MSL or within 1,200 feet of the ground regardless of MSL altitude. If the visibility is only two miles, you must stay below in Class G airspace.

You can sneak into an airport that lies in uncontrolled airspace, but if the airport has the magenta shading around it you must descend to within 700 feet of the ground while in that area. If you do this, stay away from any instrument final approach segments. You can find these in an NOS or Jeppessen approach chart book. But if the visibility is only two miles (or anything less than three but more than one), the Class E to the surface becomes a barrier. You cannot fly under Class E to the surface like you can a transition area, because the Class E to the surface touches the ground. The transition area only comes as close as 700 feet AGL. But the only vehicles operating under the Class E to the surface must be digging an underground tunnel!

A CLASS E ROADBLOCK

Under these circumstances, the Class E to the surface barrier (figure 25.1) must be avoided. The pilot must not enter the Class E to the surface for fear of an IFR/VFR collision. The pilot can fly around this zone but nowhere can she fly through the Class E to the surface. If the Class E to the surface contains the destination airport, a pilot cannot get in without a ***Special VFR Clearance.***

THE SPECIAL VFR CLEARANCE

A Special VFR may even be the savior to a really dangerous problem. If a pilot began a flight with better than basic VFR minimums but during the flight the weather got progressively worse, diversion to an airport may be the safest thing to do. If the nearest airport also has a Class E to the surface, then a Special VFR Clearance must be received before entering the zone. IFR approaches may still be in progress to the airport when the weather observer has gone home for the day, but the need for IFR/VFR separation remains. The Special VFR clearance must be issued by someone who knows if that airport will be having any IFR arrivals or departures during your proposed penetration to the airport. That person will be a controller. The controller might be at a non-radar tower, but most often he will be sitting behind a radar screen handling traffic. A Flight Service Station can get a Special VFR for a pilot, but the FSS briefer must first give a call to the controller, and the pilot gets the message third-hand.

When a Special VFR Clearance is issued, it is done only at the request of the pilot. The controller cannot say, "How about taking a Special?" If the controller does this and the pilot then has a problem, the controller is on the hook for baiting the pilot into something she might not have been prepared to handle. I heard a clever controller once say to a stranded VFR pilot, "The field is IFR due to ceilings, but another VFR pilot just landed with a Special VFR." This was the closest legal way for him to remind the pilot that asking for and receiving a Special VFR Clearance was the only way to get home.

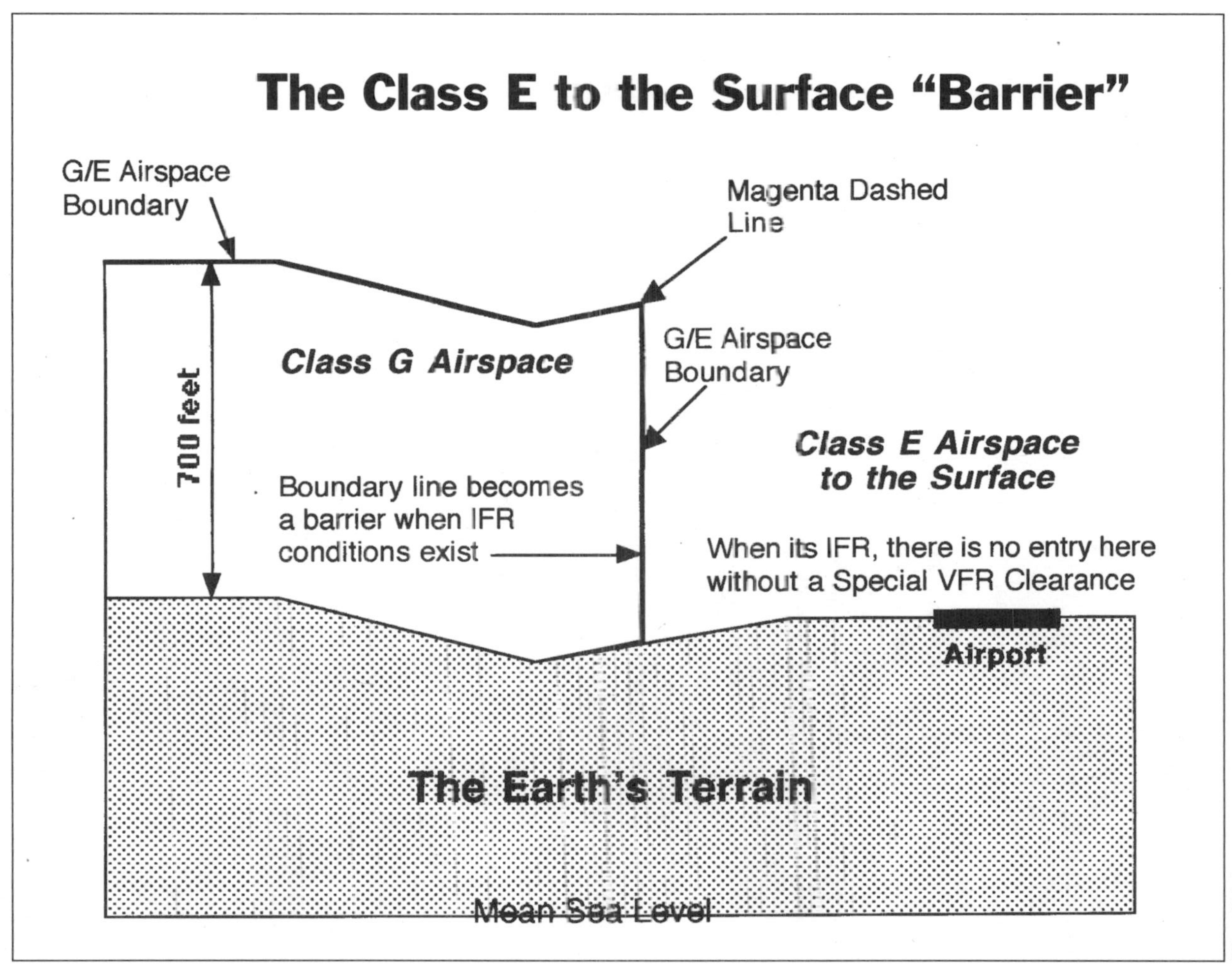
The Class E to the Surface "Barrier"
G/E Airspace Boundary
Magenta Dashed Line
Class G Airspace
700 feet
G/E Airspace Boundary
Class E Airspace to the Surface
Boundary line becomes a barrier when IFR conditions exist
When its IFR, there is no entry here without a Special VFR Clearance
Airport
The Earth's Terrain
Mean Sea Level

Fig. 25.1

A Special VFR Clearance is issued only when a collision hazard does not exist. How will the controller know this, even without radar? Because Special VFR Clearances are only issued one at a time and when there are no IFR arrivals or departures. The pilot cannot hit anybody else because there is nobody else there to hit. This holds true unless another VFR pilot violates the Class E to the surface zone. When the clearance is issued, the pilot is on his own to cross the Class E magenta dashed line and find the airport. A Special VFR Clearance can be used to depart the Class E to the surface zone when the conditions are IFR. This would allow a pilot to do touch-and-goes in the traffic pattern with less than three but more than one mile visibility. It would also allow a pilot who is headed for uncontrolled airspace to escape the Class E to get into Class G, where less than three miles but more than one mile visibility is allowed for VFR.

A Special VFR Clearance is not issued between sunset and sunrise unless the pilot has a current instrument rating and the airplane is equipped to fly IFR. Some busy airports never issue Special VFR Clearances. Airports that are very busy will have many airplanes in the Class E to the surface zone at one time. But a Special VFR requires that only a single airplane be in the zone at a time. At many large airports there is never a time, 24 hours a day, 7 days a week, 52 weeks a year, when the Class E to the surface is empty. Since the zone is never empty and since its being empty is a requirement for Special VFR, a Special VFR Clearance is never a possibility. Airports that prohibit Special VFR will have a note to this effect near the airport's data line on the chart.

When you begin comparing controlled versus uncontrolled, some confusion exists between controlled airports and controlled airspace. A controlled airport exists wherever an operating control tower is present. It is possible for an uncontrolled airport (no control tower) to be in controlled airspace if the airport has the Class E to the surface. The weather observer that is required for a Class E to the surface zone to be in effect does not have to be an FAA employee. Many airline companies will have an employee trained as a weather observer so that they can control the times when the Class E to the surface becomes effective. While in flight with visibility below three miles, it is vital that pilots stay outside controlled airspace in all its forms. Here again, the sectional chart provides the information about the horizontal and vertical boundaries of controlled airspace. The chart has clues to help pilots view it not as a flat piece of paper but as a three-dimensional Earth and Sky.

26
Class D, Class C, Class B, and Class A airspace

THIS CHAPTER WILL COVER THE "ALPHABET SOUP" CLASSES OF airspace. Classes G and E that were previously covered do not require communication with an air traffic controller, but communications are required in Classes D, C, B, and A. In addition, it is important to know how large these areas are and how to find them on a chart. Here are all the facts and figures.

Class D Airspace (figure 26.1)

>>>Class D Airspace requires communication with a controller.

>>>Class D Airspace contains a primary airport and that airport has an Air traffic control tower.

>>>Class D Airspace may not have RADAR control, therefore traffic advisories are not provided.

>>>Class D Airspace requires 3 statute miles of visibility.

>>>Class D Airspace requires that VFR flights maintain the minimum distance from clouds of 500' below, 1,000' above, and 2,000' horizontal.

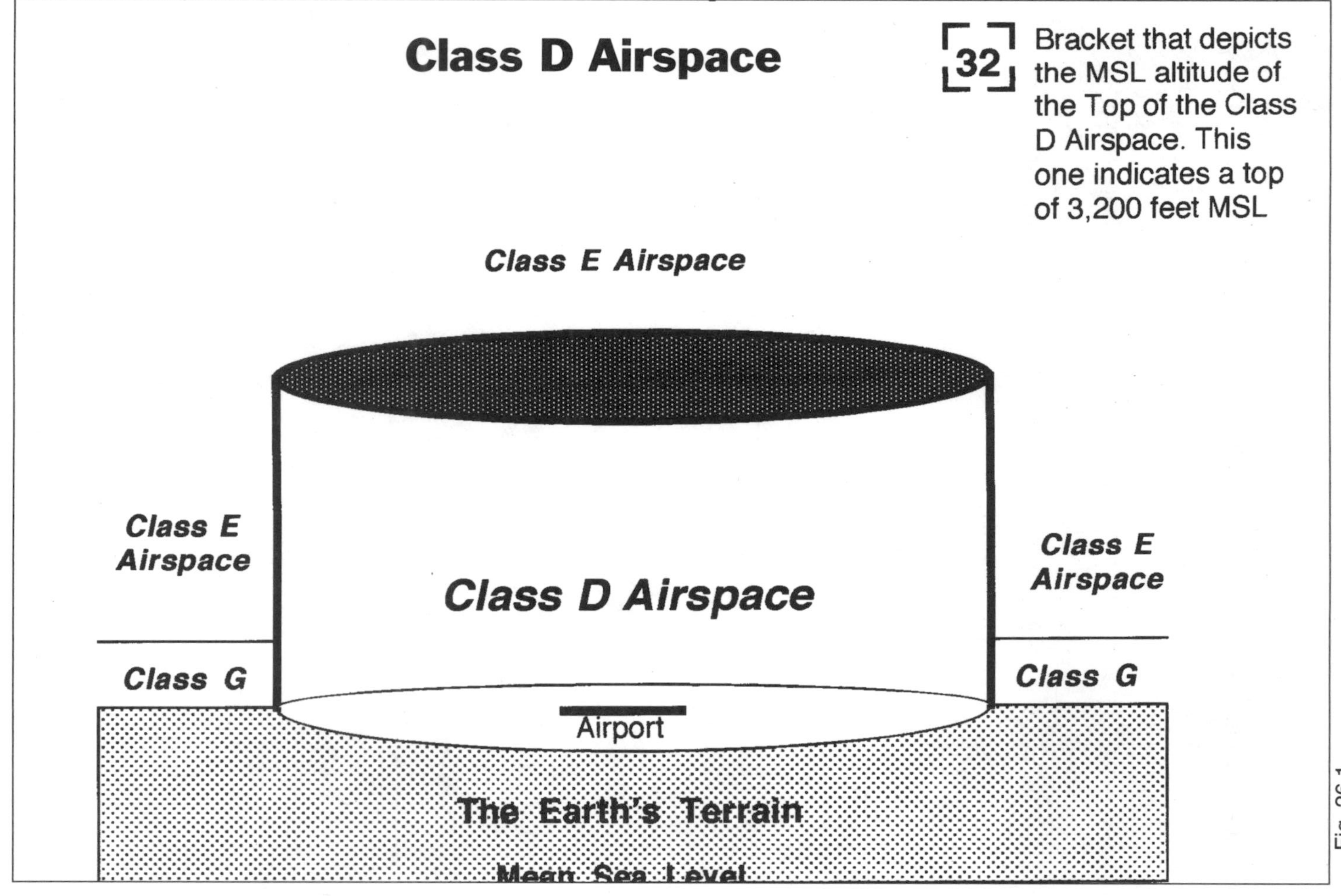
Class D Airspace
32
Bracket that depicts the MSL altitude of the Top of the Class D Airspace. This one indicates a top of 3,200 feet MSL
Class E Airspace
Class E Airspace
Class D Airspace
Class E Airspace
Class G
Class G
Airport
The Earth's Terrain
Mean Sea Level

Fig. 26.1

>>>Class D Airspace allows flight by Student Pilots.

>>>Class D Airspace does not require a Transponder.

>>>Class D Airspace is "cylinder" shaped. The top of Class D Airspace is usually 2,500' AGL. The equivalent MSL altitude is shown in a bracket on the chart. The base is 5 statute miles in radius.

>>>In Class D Airspace a Special VFR Clearance can be obtained.

>>>Class D Airspace is depicted on the chart with a blue dashed line.

Class C Airspace (figure 26.2)

>>>Class C Airspace requires communication with a controller before entering the Outer Circle. The controller must say your airplane's call sign prior to entry.

>>>Class C Airspace contains a primary airport and that airport has an air traffic control tower.

>>>Class C Airspace has RADAR control that provides separation between IFR and VFR traffic.

>>>Class C Airspace requires 3 statute miles of visibility.

>>>Class C Airspace requires that VFR flights maintain the minimum distance from clouds of 500' below, 1,000' above, and 2,000' horizontal.

>>>Class C Airspace allows flight by Student Pilots.

>>>Certain Class C's may allow Special VFR Clearances. If a Class C does not accept a Special VFR Clearance, a note will say "NoSVFR" on the chart.

>>>Class C Airspace requires an Altitude Reporting (Mode C) Transponder to operate within or above the Class C Inner and Outer Circles.

>>>Class C Airspace is "upside-down wedding cake" shaped. The top of Class C Airspace is usually 4,000' AGL. The equivalent MSL altitude is shown on the chart. The Inner Circle is 5 nautical miles in radius. The Outer Circle is 10 nautical miles in radius, and the Outer Area is 20 nautical miles in radius.

>>>The Inner and Outer Circles of Class C Airspace are depicted on the chart with solid magenta lines. The Outer Area is not depicted.

Class B Airspace (figure 26.3)

>>>Class B Airspace requires communication with a controller. The controller must say "You are cleared to enter the Class B Airspace."

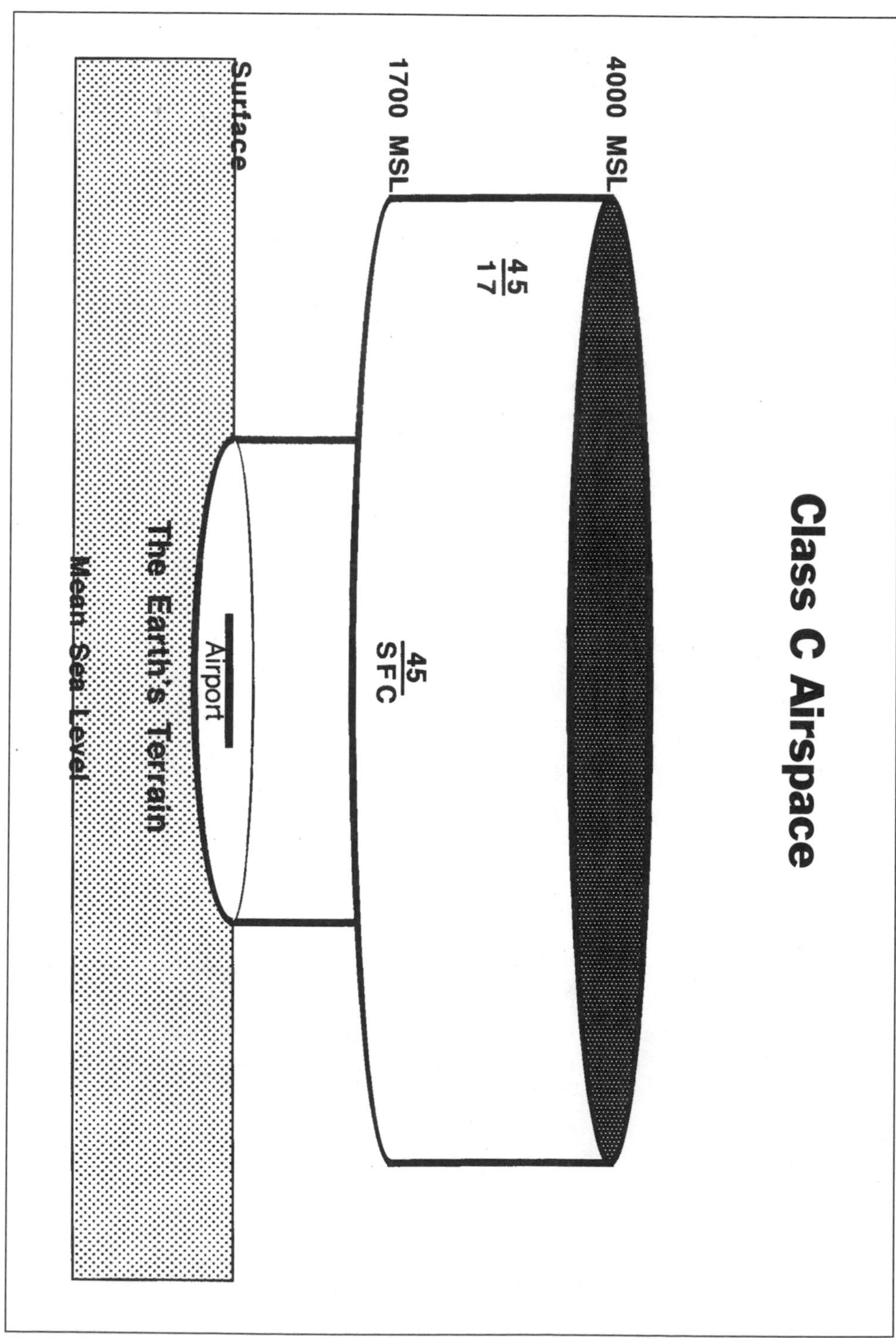
Class C Airspace
4000 MSL
1700 MSL
Surface
45
17
45
SFC
Airport
The Earth's Terrain
Mean Sea Level

Fig. 26.2

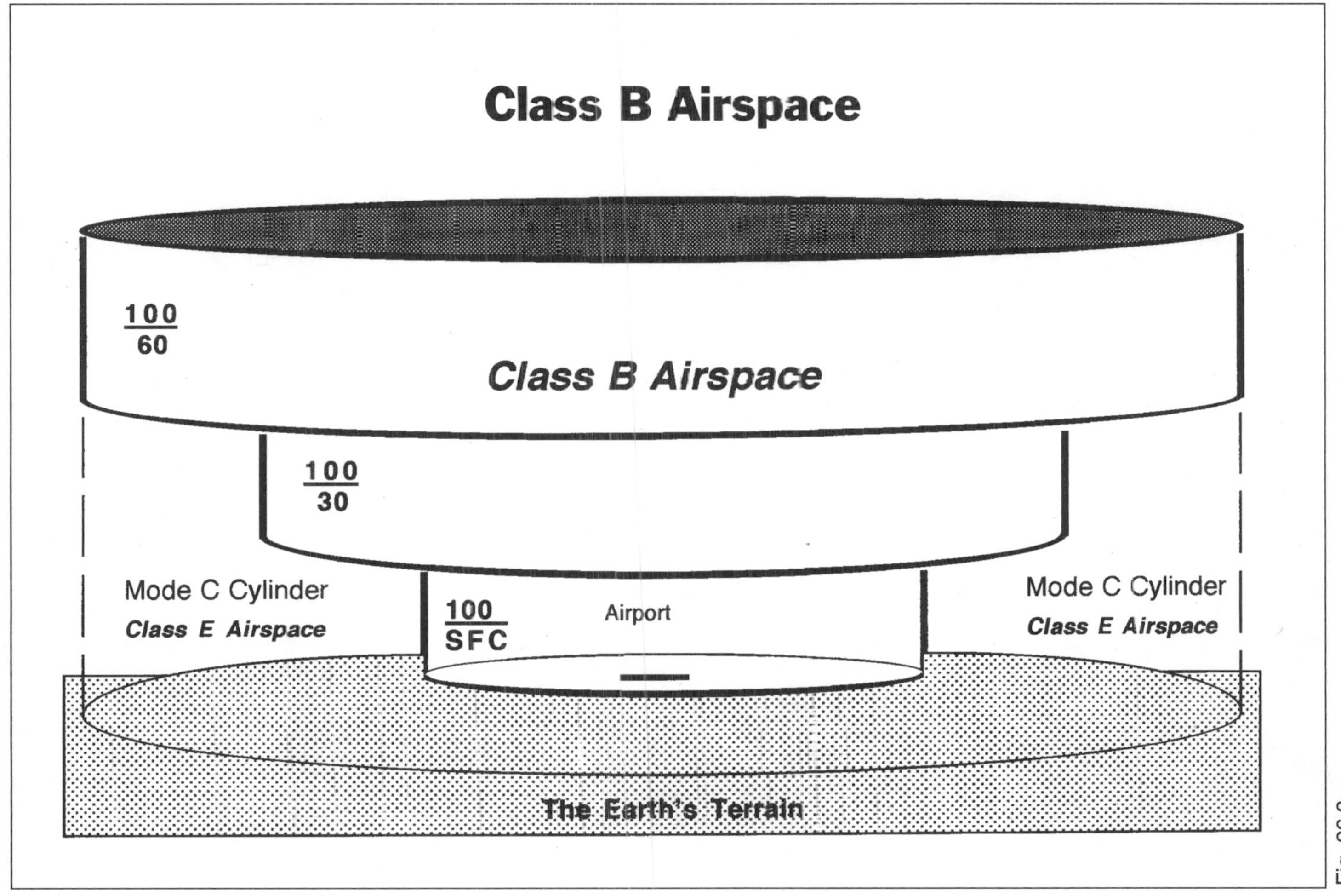

Class B Airspace
100
60
Class B Airspace
100
30
Mode C Cylinder
Class E Airspace
100
SFC
Airport
Mode C Cylinder
Class E Airspace
The Earth's Terrain

Fig. 26.3

>>>Class B Airspace contains a primary airport and that airport has an air traffic control tower.

>>>Class B Airspace has RADAR control that provides separation between IFR and VFR traffic.

>>>Class B Airspace requires 3 statute miles of visibility.

>>>Class B Airspace requires that VFR flights maintain clear of the clouds.

>>>Certain Class B's do not allow flights by Student Pilots. The Student Pilot prohibited airport list is written in FAR 91.131. Student Pilots are permitted in all other Class B's with the proper instructor endorsement.

>>>Certain Class B's may allow Special VFR Clearances. If a Class B does not accept a Special VFR Clearance, a note will say "NoSVFR" on the chart.

>>>Class B Airspace requires an Altitude Reporting (Mode C) Transponder to operate within or above a "cylinder" shape that has a 30-nautical-mile radius. The 30-mile Mode C ring is indicated on the chart with a thin, solid, blue line.

>>>Class B Airspace is "upside-down wedding cake" shaped with many tiers. The top of Class B Airspace is depicted in the chart in MSL feet and the complete shape of this airspace will vary from one Class B to the next.

>>>Class B Airspace is depicted on the chart with solid blue lines at each boundary ring. This line is slightly thicker than the Mode C line.

Class A Airspace

>>>Class A Airspace starts at 18,000 feet MSL and extends to 60,000 feet MSL.

>>>Only IFR flights are allowed in Class A Airspace.

>>>Class A Airspace is not depicted on the chart.

27
The "vertical" airspace diagram

CHAPTER 26 WAS SIMPLY A LIST OF THE REQUIREMENTS AND dimensions of the various airspace types. Your understanding of airspace must go beyond just memorizing the lists of Chapter 26. We fly in a three-dimensional sky, and the airspace types are also three dimensional. The sectional charts that we use to identify the airspace is flat and the uninformed pilot sees the chart as flat with only two dimensions: North-South or East West. But real pilots see the third dimension: which is "up!" This chapter will help you see the third dimension. In this chapter we look at airspace types as they exist "above" a certain location on the chart. In this chapter we draw the "vertical" airspace diagram. The diagrams shown all stop at 18,000 feet MSL. Above 18,000 feet and continuing up to 60,000 feet is Class A airspace. Only IFR flights are allowed in Class A. Our discussion is limited to VFR flights so that means that the highest we can fly is 18,000 feet.

LOOKING UP FOR AIRSPACE TYPES

Figure 27.1 is a single tower off in the middle of nowhere. The figure is small, but there is no magenta shading around the symbol of the tower. The chart shows us what is North, East, South, and West of the tower, but what about the world that exists *above* the tower? Figure 27.1 is a diagram of the airspace that exists over the tower. If you launched a rocket beside the tower and the rocket flew straight up, what airspace types would it pass through on its way to orbit? To answer this question you must determine the elevation of the ground.

The tower symbols hold the key to discovering the ground elevation. The tower has a bold, blue number 708 beside the tower symbol. The number 708 is the feet above sea level of the top of the tower. The tower symbol also has a number, 204, in parentheses. The number 204 is the feet the top of the tower is above the ground. If the top of the tower is 708 MSL and at the same time 204 AGL, then the ground's elevation must be the difference between the two, or 504. In figure 27.1 the ground's elevation is shown at 504.

Once the elevation of the ground is known, the overlying airspace can be determined. The tower is in an area where there is no magenta shading, therefore uncontrolled airspace (Class G) is on the surface and is 1,200 feet thick. In other words, the Class G and Class E boundary is 1,200 feet higher than the ground. Since the ground is already 504 feet above sea level, and then another 1,200 feet is added above that to reach the boundary, then the boundary is located at 1,704 feet MSL (504 + 1,200 = 1,704). The G/E boundary is indicated on figure 27.1.

In this location, above the tower, between 1,704 MSL and 18,000 feet MSL is all Class E airspace. At 10,000 feet MSL a Mode C (altitude reporting) transponder is required.

Figure 27.2 depicts the Washington County Airport. What is the "vertical" airspace above this airport? The chart indicates the ground elevation, in the airport information line, to be 955 feet above sea level. Now that the ground's elevation is known, we ask the question, *"Where is the controlled airspace above this airport?"* There is no magenta shading, so the G/E boundary is 1,200 feet higher than the ground. The ground is already 955, so the boundary must be 2,155 feet MSL (955 + 1,200 = 2,155). Between 2,155 feet MSL and 18,000 feet MSL is Class E airspace. Mode C required above 10,000 MSL.

Figure 27.3 is the Walker County - Bevill Airport. This airport is 482 feet above sea level, but this airport is surrounded by a magenta shaded circle. Everywhere within that circle the controlled airspace (Class E) dips down to 700 feet of the surface. Figure 27.3 indicates the airport elevation of 482, and then the 700-foot gap between the Earth and the G/E boundary line (482 + 700 = 1,182 feet MSL). Figure 27.4 shows the Crossville Memorial Airport. This airport has an elevation of 1,881. (*Note:* The ground is higher at Crossville than the G/E boundary was over Walker County Airport in figure 27.3.) The airspace symbols are different at Crossville than we have seen before. There are two circles surrounding Crossville. One circle is of magenta shading, the other is a magenta dashed line. Inside the dashed line, the controlled airspace (Class E) comes all the way down and touches the surface. Figure 27.4 indicates the airport elevation of 1,881, but Class E airspace starts right on the ground and continues up to 18,000 feet MSL. A Special VFR would be required to land at Crossville when VFR weather minimums did not exist.

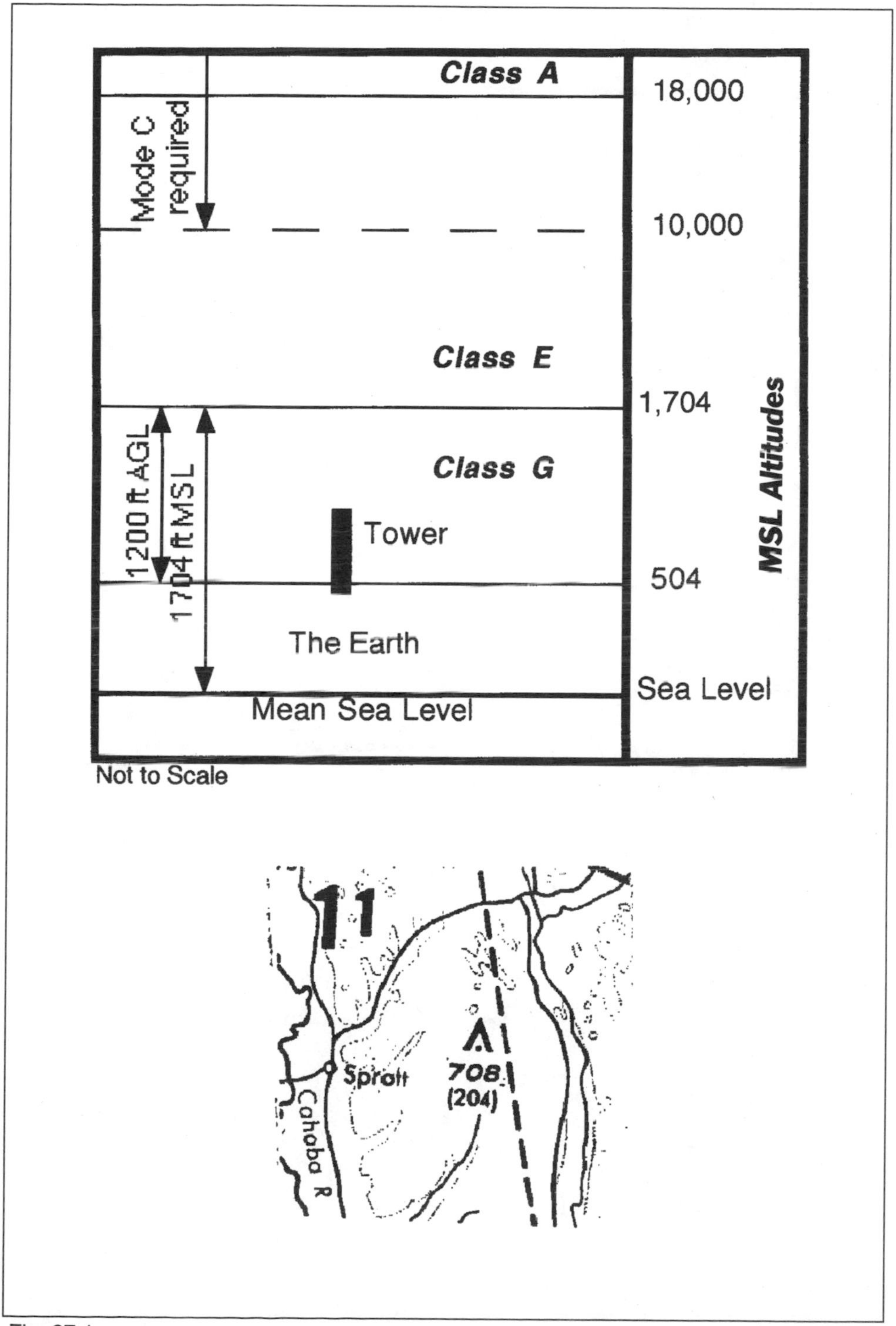
Class A
18,000
Mode C required
10,000
Class E
1,704
MSL Altitudes
1200 ft AGL
1704 ft MSL
Class G
Tower
504
The Earth
Sea Level
Mean Sea Level
Not to Scale
11
Sprott
708
(204)
Cahaba R

Fig. 27.1

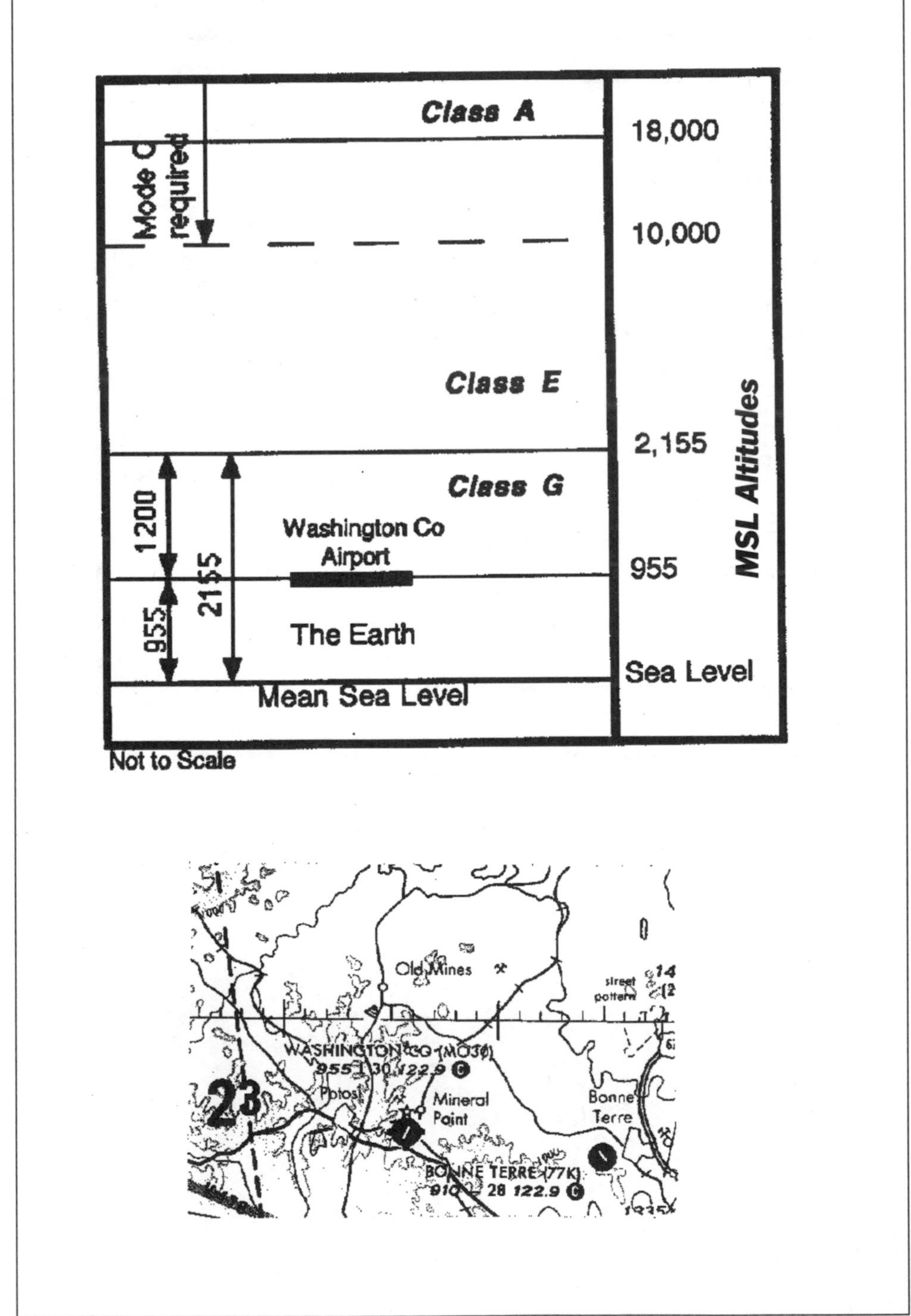
Class A
18,000
Mode C required
10,000
Class E
2,155
MSL Altitudes
Class G
1200
Washington Co Airport
955
2155
955
The Earth
Sea Level
Mean Sea Level
Not to Scale
Old Mines
street pattern
WASHINGTON CO (MO30)
955 L 30 122.9
23
Potosi
Mineral Point
Bonne Terre
BONNE TERRE (77K)
910 28 122.9

Fig. 27.2

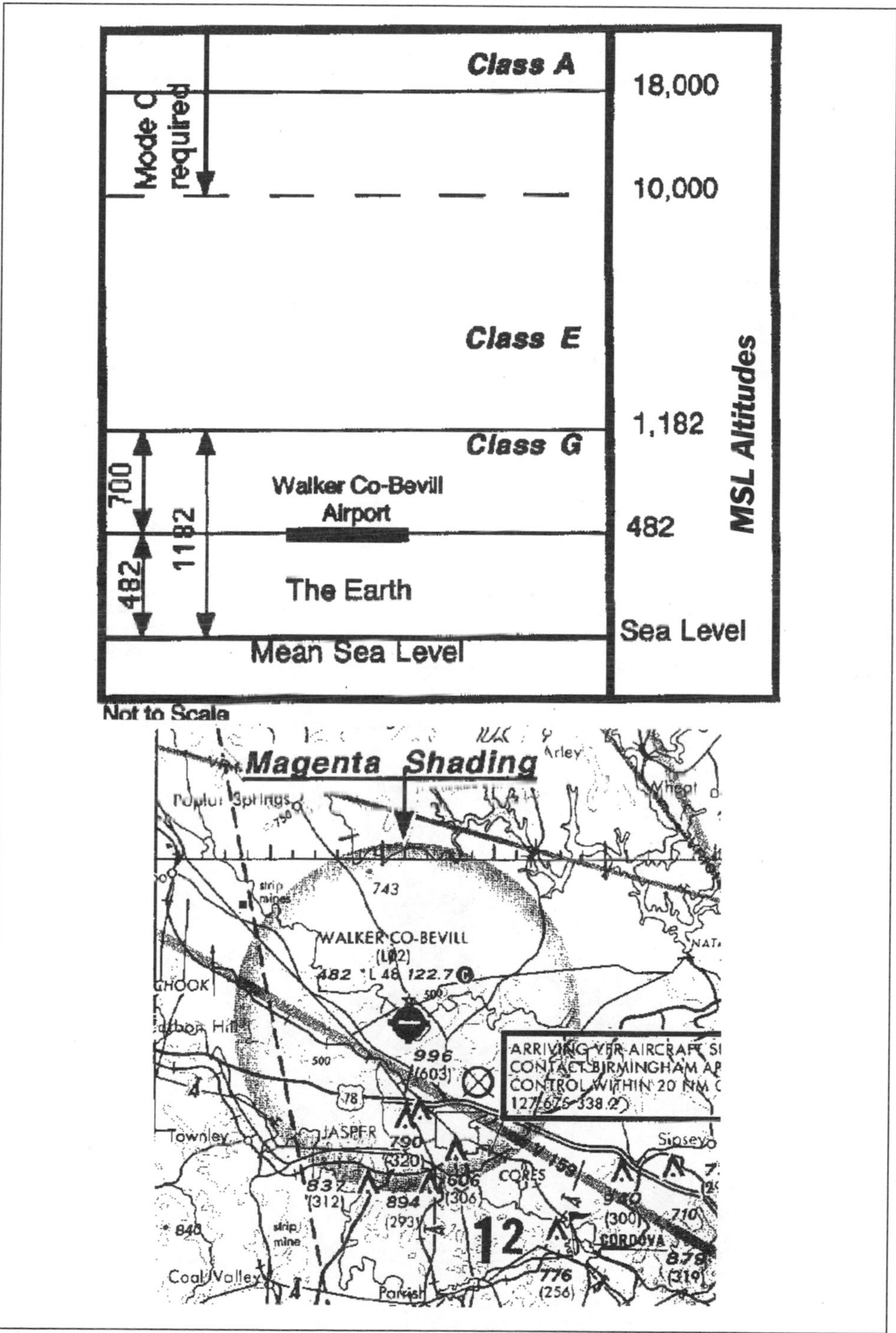

Class A
18,000
Mode C required
10,000
Class E
MSL Altitudes
1,182
Class G
700
1182
Walker Co-Bevill Airport
482
482
The Earth
Sea Level
Mean Sea Level
Not to Scale
Magenta Shading
Arley
743
WALKER CO-BEVILL
482
122.7
996
(603)
ARRIVING VFR AIRCRAFT
JASPER
790
(320)
837
(312)
894
(293)
606
(306)
12
CORDOVA
776
(256)
Coal Valley
Townley
Sipsey

Fig. 27.3

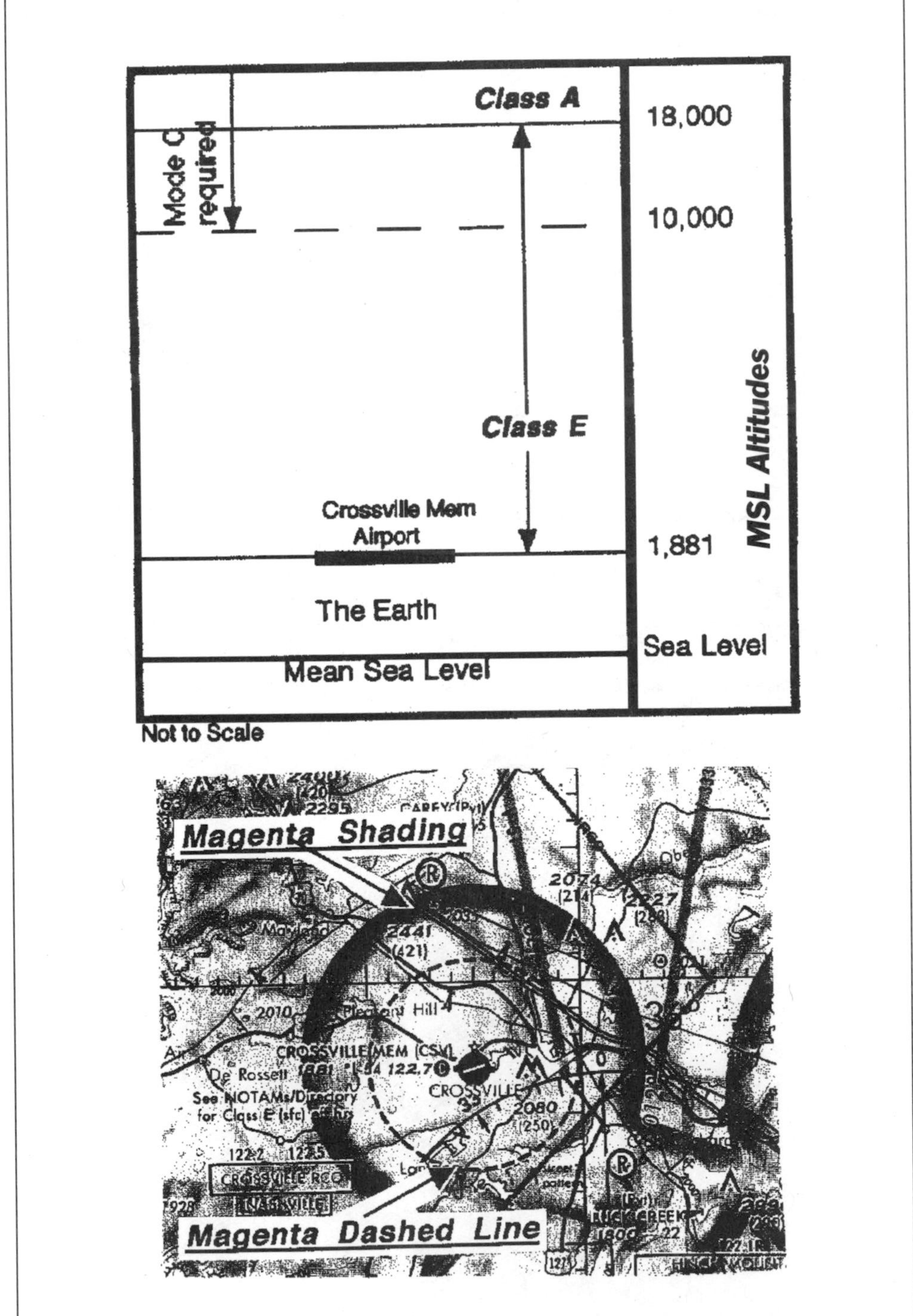
Class A
18,000
Mode C required
10,000
Class E
MSL Altitudes
Crossville Mem Airport
1,881
The Earth
Sea Level
Mean Sea Level
Not to Scale
Magenta Shading
2441
(421)
2074
(214)
2010
Pleasant Hill
CROSSVILLE MEM (CSV)
De Rossett
See NOTAMs/Directory for Class E (sfc) eff hrs
CROSSVILLE
2080
(250)
122.2
122.5
CROSSVILLE RCO
NASHVILLE
Magenta Dashed Line

Fig. 27.4

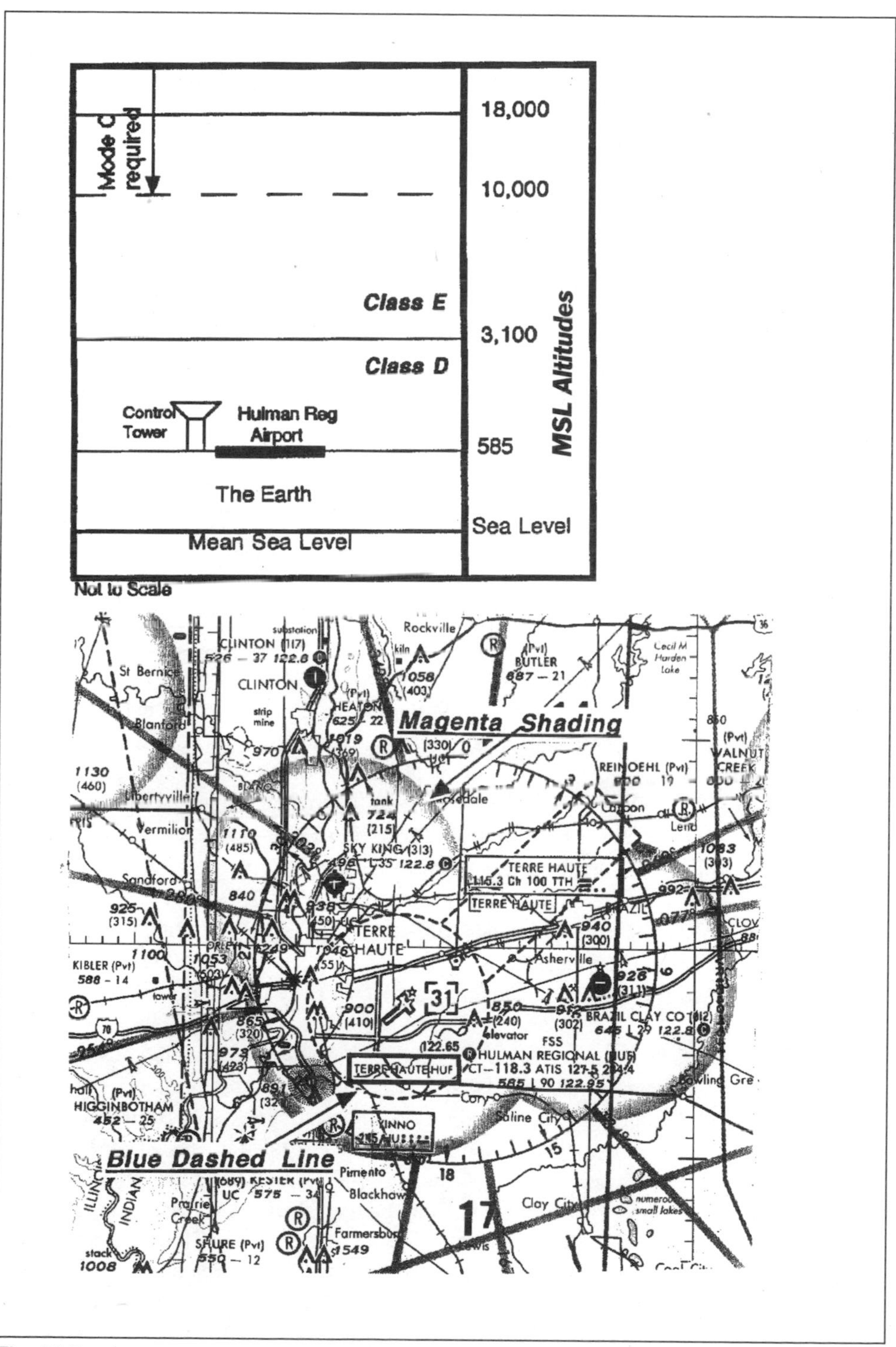

Mode C required
18,000
10,000
Class E
3,100
Class D
Control Tower
Hulman Reg Airport
585
MSL Altitudes
The Earth
Sea Level
Mean Sea Level
Not to Scale
Magenta Shading
Blue Dashed Line
Rockville
CLINTON
TERRE HAUTE
HULMAN REGIONAL
Saline City
Clay City
Pimento
Blackhawk
Farmersburg

Fig. 27.5

Figure 27.5 is the Hulman Regional Airport located at an elevation of 585 feet MSL. This airport is surrounded by a blue dashed line and has a blue airport symbol. The blue indicates that there is a control tower present on the airport and that means that a Class D airspace type is present. Figure 27.5 illustrates the ground at 585 and a Class D sitting on the ground. The Class D extends up to 3,100 feet MSL. We know how high the Class D airspace goes because of the symbol that is just to the right of the airport symbol on the chart. The number 31 is contained in a four-sided bracket. This number is the "hundreds" of feet above sea level of the top of Class D at this location. Other Class D's will have different top altitudes. How thick is this Class D? Well, the top is at 3,100 feet MSL and the bottom (the Earth) is at 585 feet MSL, so it is 2,515 feet thick. Above 3,100 MSL is Class E. Class E extends all the way up to Class A at 18,000.

Figure 27.6 is the area around Chattanooga, Tennessee. The central airport of the diagram is Lovell Airport, which has a Class C airspace type. Just East of Lovell by 9 miles is the Collegedale Airport. The vertical diagram of figure 27.6 is of the airspace above Collegedale. Collegedale has an elevation of 860 and is within both a magenta shading and a magenta solid line. As always, the magenta shading indicates the G/E boundary line to be 700 feet above the ground, which would be 1,560 feet MSL in this case (860 + 700 = 1,560). Figure 27.6 indicates that Class G airspace is on the surface at Collegedale but that Class E begins at 1,560. But Collegedale is also within the solid magenta lines of Lovell's Class C airspace. Just to the Northwest of the Collegedale Airport symbol is a Class C airspace altitude symbol. It shows the number 47 above a line and the number 26 below the line. These numbers are the "hundreds" of MSL feet that make up the top and bottom of the Class C airspace. Over Collegedale the base of Class C airspace is shown in Figure 27.6 to be at 2,600 feet MSL and the top of Class C to be at 4,700 feet MSL. Mode C (altitude reporting) is required anytime an airplane is with Class C or above Class C airspace. Above 4,700 MSL is Class E again and this extends up to Class A at 18,000 feet MSL.

Figure 27.7 is the area around Charlotte, North Carolina, which has a Class B airspace. Located at 34° 59' by 81° 04' (Southwest of Charlotte) is the Rock Hill-York County Airport. This airport underlies the Class B airspace, but is also surrounded by magenta shading. Figure 27.7 illustrates the airport elevation of 669. Class G goes from the surface to 700 feet above the surface within the magenta shading, therefore the G/E boundary is located at 1,369 feet MSL (669 + 700 = 1,369). Class E goes from 1,369 up to the base of Class B airspace. The base and top of the Class B airspace is indicated in the area above Rock Hill to be 3,600 MSL for the base and 10,000 MSL for the top. The base and top airspace symbol is located Northwest of Rock Hill. Above the Class B at 10,000 is

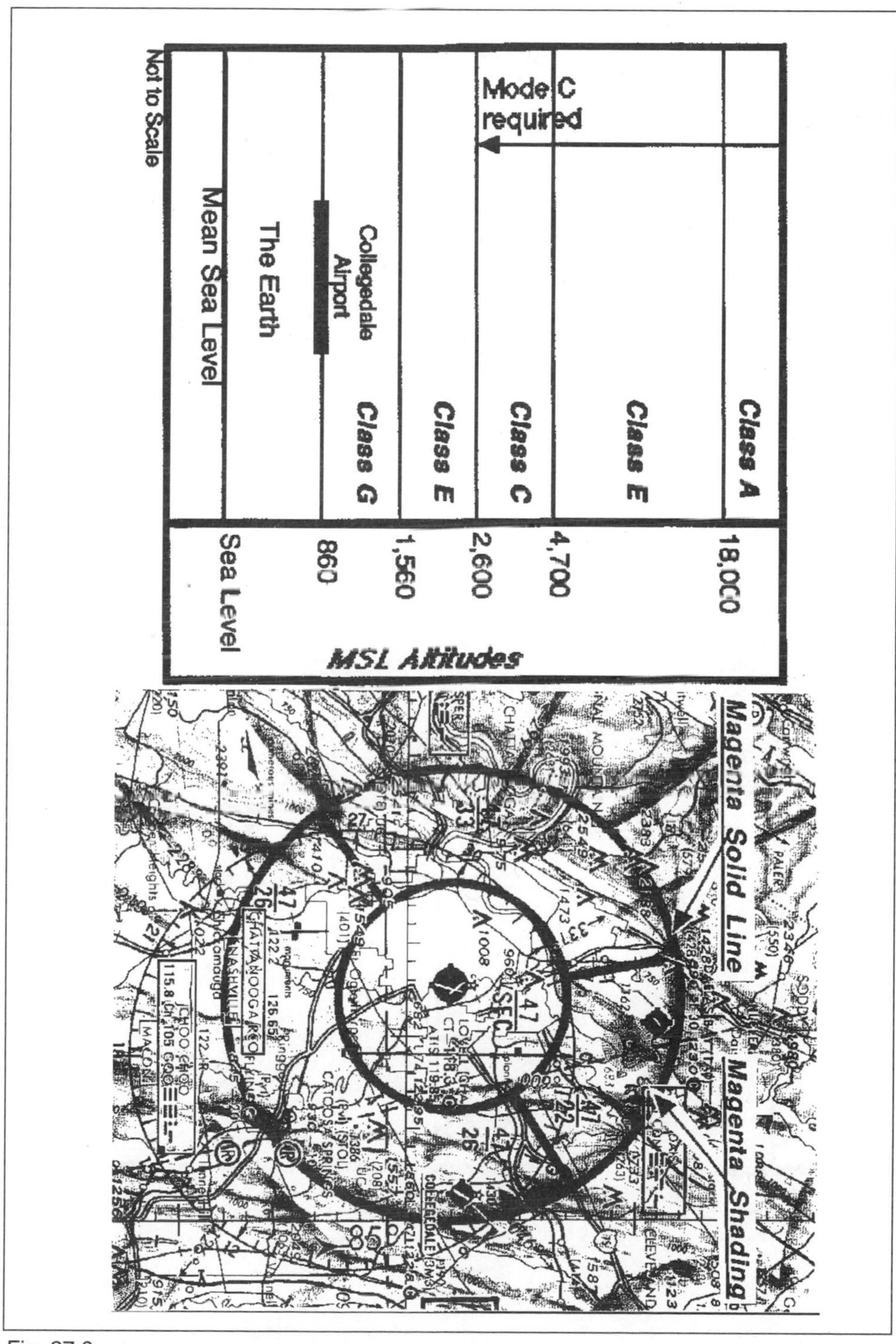
Class A
Class E
Class C
Class E
Class G
18,000
4,700
2,600
1,560
860
Sea Level
MSL Altitudes
Mode C required
Collegedale Airport
The Earth
Mean Sea Level
Not to Scale
Magenta Solid Line
Magenta Shading

Fig. 27.6

Class E, which goes up to the Class A at 18,000 MSL. The Rock Hill Airport is also surrounded by a thin, solid, blue, Mode C Ring. This means that a Mode C (altitude reporting) transponder is required at any altitude within the ring.

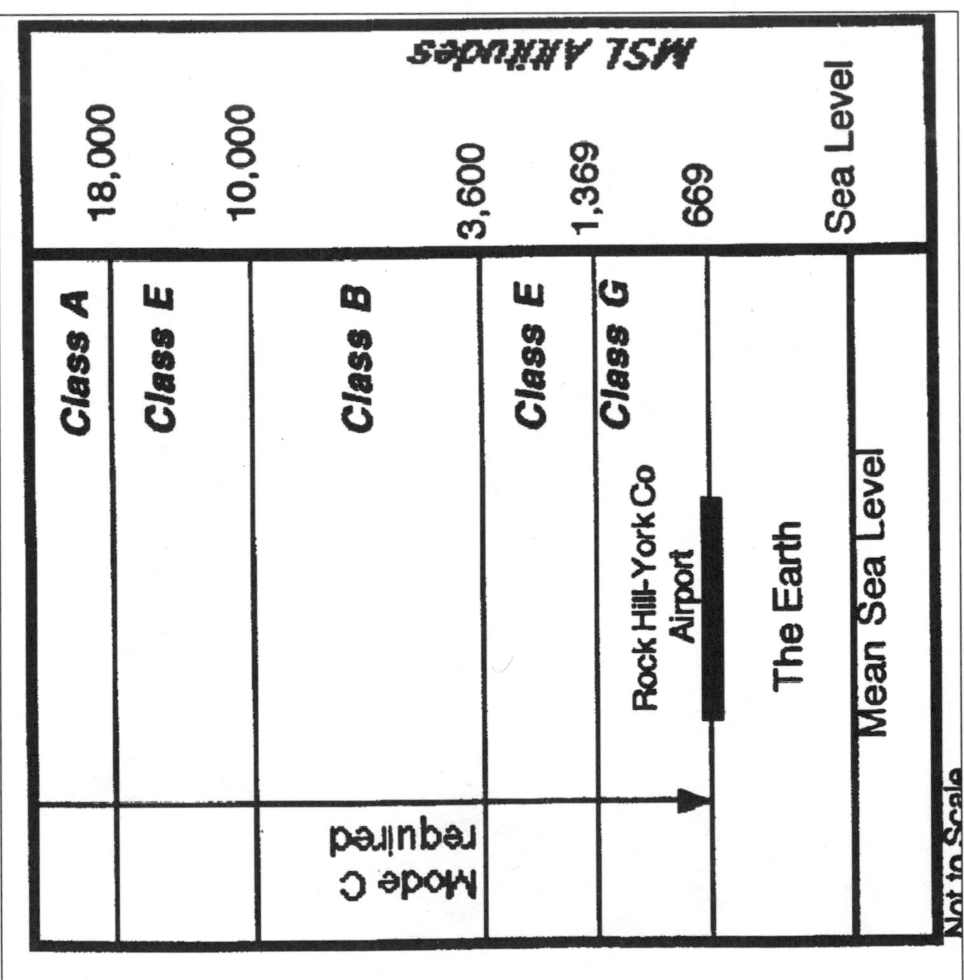

Fig. 27.7a

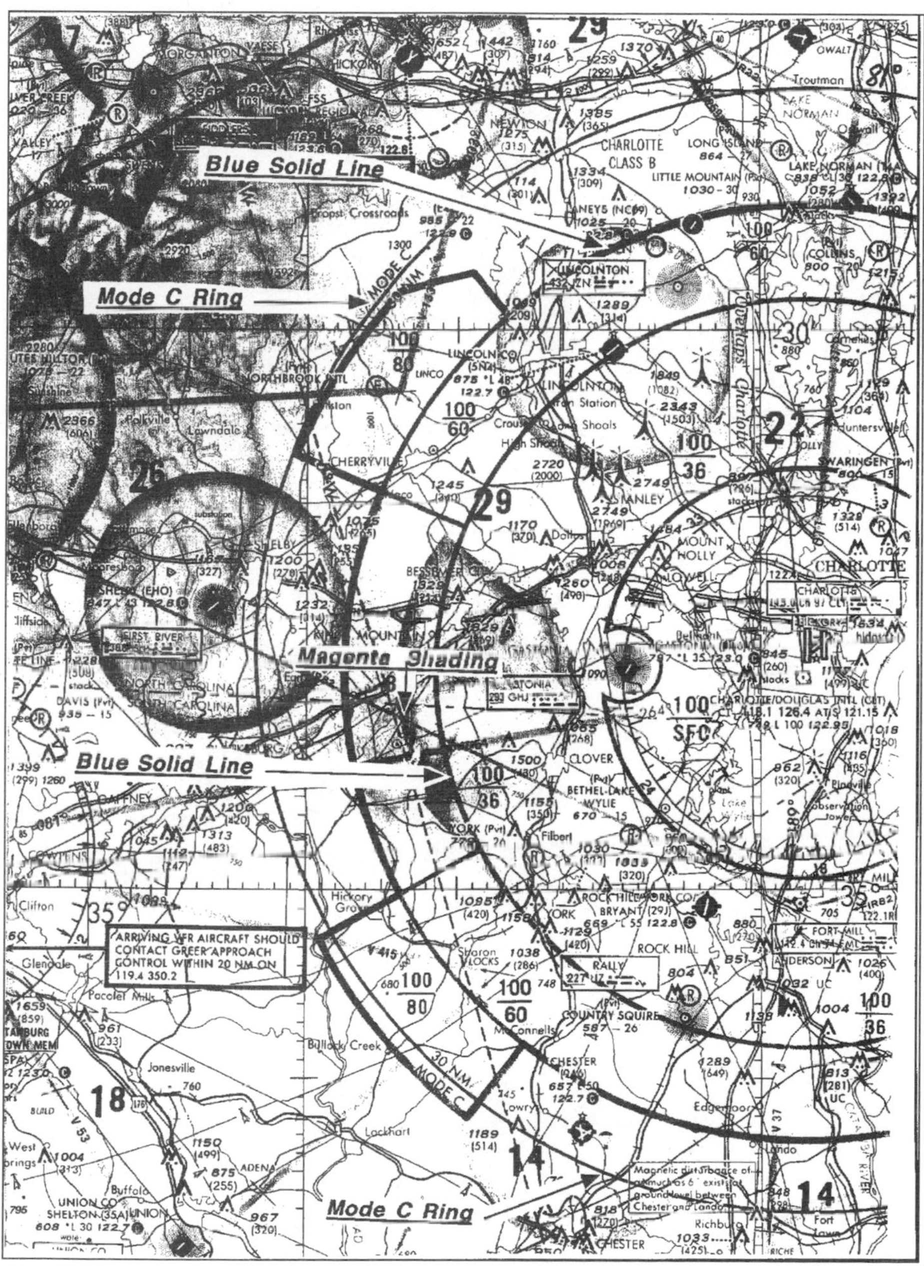
Blue Solid Line
CHARLOTTE
CLASS B
Mode C Ring
Magenta Shading
Blue Solid Line
ARRIVING VFR AIRCRAFT SHOULD CONTACT GREER APPROACH CONTROL WITHIN 20 NM ON 119.4 350.2
Mode C Ring

Fig. 27.7b

28
Special use airspace

SOME AIRSPACE HAS BEEN ALLOCATED TO THE MILITARY FOR TRAINING purposes. These areas range from very dangerous to civilian traffic to harmless. The danger might change from day to day, or be based on time of day. Consider a scale from 1 to 10 where 10 is "no trespassing" and 1 is "come on in, but watch out." We will discuss Prohibited Areas, Restricted Areas, Temporary Restricted Areas, Warning Areas, Alert Areas, Military Operations Areas, Military Training Routes, and Controlled Firing Areas to determine where on the 1 to 10 scale they belong.

PROHIBITED AREA

Prohibited Areas (blue chart boundary) are a 10. Civilian flights are always prohibited in Prohibited Areas. The most conspicuous Prohibited Area is the Mall in Washington, D.C. Within this area lies the United States Capitol, the White House, and the Monuments. (We know now that this Prohibited Area can be penetrated at least by a man in a Cessna willing to kill himself!) The National Air and Space Museum ironically is within a no-flying zone.

There is a moving Prohibited Area around the President and Vice President at all times. Even former presidents have Prohibited Areas over their residences. There are circular Prohibited Areas across the American midwest, each covering an Intercontinental Ballistic Missile silo. Temporary Prohibited Areas can be formed by NOTAMs whenever the FAA Administrator decides to do so. Temporary Prohibited Areas are formed during Space Shuttle launches and recoveries. Once a Prohibited Area was thrown up around the location where the top secret Stealth fighter crashed so that no one could get a look at it.

RESTRICTED AREA

Restricted Areas (blue chart boundary) are a 9. Most of the time, flying a civilian airplane into a "hot" Restricted Area would be life threatening. The military is lobbing artillery shells through the air, or firing at drones, or dropping bombs. The reason these areas are classified as Restricted rather than Prohibited is that there are times when these areas go "cold."

Cold means inactive. Hot means stay out. On the lower tab of sectional charts a white strip can be found. That strip contains information concerning all Special Use Airspace on that chart. A Restricted Area's time of operation will be indicated here. Some Restricted Areas are hot all the time; others have a schedule of operation. Before flying into any Restricted Area, radio the controlling agency to determine the status of the area. Do not even trust NOTAMs. The time of the NOTAM may change without notification. If an IFR flight plan is filed with a route of flight through a Restricted Area and the plan is approved as filed, then no further clearance will be required. In that case, the Air Traffic Control computer will have verified that the Restricted Area is cold and that you are cleared to pass through.

WARNING AREA

Warning Areas (blue chart boundary) are a 9. The danger to civilian aircraft in a Warning Area is identical to the dangers inside a Restricted Area. So why do they go by separate names? It is a matter of airspace ownership. Warning Areas are over international waters that begin beyond the three-mile offshore limit. Airspace over international water is considered international airspace. The United States cannot fully restrict international airspace because it is not ours to restrict. Instead, the military merely warns pilots to stay away. It is a warning worth heeding!

ALERT AREA

Alert Areas (blue chart boundary) are a 5. These areas contain a high level of military flying activity; no bombing or firing, just lots of fast moving aircraft. Civilian pilots do not need special authorization to fly within an Alert Area, but extreme caution is urged.

MOAs

Military Operations Areas (MOAs — magenta chart boundary) are a 5. In comparison, civilian flight schools usually set up unofficial "practice areas" for students to perfect flight maneuvers. MOAs are "practice areas" for military training. Alert Areas are usually situated around military bases, while MOAs are out in uncongested areas. Civilian pilots are encouraged to use caution when flying within an MOA, but special clearance is not required. Many MOAs do not reach down to the ground. An MOA can be a block of airspace that starts at a predetermined MSL altitude. Check the tab at the bottom of the sectional chart for MOA dimensions.

MTRs

Military Training Routes (MTRs) are a 4. Thin gray lines on a sectional chart indicate preset military routes. The routes are identified by either a VR or IR designation. These routes can be flown under IFR conditions where the IR designation exists, or under VFR conditions. If the route has a four digit identification number (example: IR 2314) then this indicates that the route will be flown at or below 1,500 feet AGL. Identification numbers with two or three digits may be above or below 1,500 AGL. These routes are for high speed, low altitude practice runs. Military jets pay no attention to civilian speed limits, so it is not uncommon for these aircraft to do better than 400 knots at treetop level. Some military wings are good about calling the local FSS and announcing the use of a particular route. Some wings are not. The information from the FSS about hot or cold MTRs can be totally unreliable. The best idea is to avoid the area altogether. The rating here was a 4. This was lower than MOAs and Alert Areas because pilots should be able to narrow down the location of this traffic by observing the path the gray lines take. But, the gray lines are a misconception as well. They look very thin on the chart, but the jets can spread out on either side of the lines for several miles. Another clue: If you see one military jet, look for another one quickly. They usually travel in pairs.

CONTROLLED FIRING AREA

Controlled Firing Areas are not shown on either VFR or IFR charts. The military conducts artillery firing that would be quite hazardous to civilian airplanes, yet the military does not tell us where these areas are located! Theoretically when a civilian airplane flies into these areas, the military will stop firing until the airplane passes. They say they detect the airplane's approach with radar, spotter planes, and ground observers. I am not convinced! This one is quite scary and its also impossible to place on the scale from 1 to 10 because we never know where these areas exist!

Glossary

Agonic line The line where there is zero magnetic variation. This line runs through both the Magnetic and True North Poles and then extends to the equator.

Airmet A hazardous weather advisory specifically for light aircraft. Usually moderate icing, moderate turbulence, or widespread areas of low IFR conditions will trigger an Airmet.

Alert Area An area around a military airport where a high level of aircraft activity can be expected.

Attitude Direction Indicator (ADI) An electronic instrument that displays on a CRT screen the attitude of the aircraft as well as information on glide slope and localizer position.

Automatic Direction Finder The instrument that receives the signal from a Nondirectional Beacon (NDB) and displays the NDB's position relative to the nose of the aircraft.

Calibrated Airspeed The Indicated Airspeed corrected for the "installation error." See chapter 7.

Class A Airspace The airspace at and above 18,000 feet MSL and extending to 60,000 feet MSL. Class A requires an IFR Flight Plan and an IFR Clearance.

Class B Airspace The airspace around America's busiest airports. Flight into a Class B requires radio communications and a clearance. Class B also requires an altitude reporting, Mode C transponder.

Class C Airspace The airspace around America's mid-sized airports. Flight into a Class C requires radio communications and an altitude reporting Mode C transponder.

Class D Airspace The airspace around an airport with an operating control tower. Flight into a Class D requires radio communications.

Class E Airspace Controlled airspace where certain visibility and distance from clouds must be maintained. See chapter 23.

Class G Airspace Uncontrolled airspace where certain visibility and distance from clouds must be maintained. See chapter 23.

Climb Performance Chart A chart, usually provided by an aircraft manufacturer, that predicted the aircraft's time to climb, distance traveled while climbing, and fuel burned during the climb.

Compass Correction Card The card, mounted inside the aircraft, that corrects for Deviation errors in the aircraft due to electrical equipment and radios.

Contour lines The lines on a chart that depict elevation changes.

Control Display Unit (CDU) The device that allows a pilot to interact with a Flight Management Computer. See chapter 22.

Dead Reckoning The form of navigation where a pilot uses a True Course, then applies corrections for wind, magnetic variation, and aircraft deviation to arrive at a Compass Heading to be used on a cross-country flight.

Deviation The errors of the magnetic compass that are caused by electronic equipment and radios within the aircraft.

Differential GPS The use of the Global Positioning System together with a ground-based antenna to increase navigational accuracy. DGPS is used for precision instrument approaches.

Direct User Access Terminal (DUAT) The computer system that passes weather information from observers and forecasters to pilots.

Electronic Flight Instrument System (EFIS) The system that encompasses the Electronic Attitude Direction Indicator and the Electronic Horizontal Situation Indicator. See chapter 22.

Electronic Horizontal Situation Indicator (EHSI) The screen that shows the pilot an overhead display of the airplane and its surroundings. The display can depict the aircraft heading, ground track, and precipitation echoes.

Eratosthenes The man whose calculations, made in 240 B.C., proved with the use of shadow angles, that the Earth was round.

Equator The line of Zero degrees Latitude. It is equidistant from the True North and the True South Poles.

Flight Computer A pilot's manual navigation tool used to calculate True Airspeed, Groundspeed, Fuel Consumption, and many other calculations. The Flight Computer has a "calculator" side and a "wind face" side.

Flight Management Computer (FMC) An onboard computer that monitors the navigation and performance of an aircraft. The FMC is accessed by the pilot using the Control Display Unit.

Flight Management System The integration of many electronic systems, including the auto-throttles, the navigation data bases, the autopilot, and the Flight Management Computer.

Flight Service Station (FSS) The FAA station where weather information is observed, forecasted, and disseminated to pilots. The FSS also records, activates, and closes VFR flight plans.

Flight Watch The "enroute weather specialists" at particular stations around the country. Flight Watch (also known as Enroute Flight Advisory Service - EFAS) can be reached at 122.0.

Fuel Consumption table The table, usually provided by an aircraft manufacturer, that predicts fuel consumption during various conditions of altitude and temperature.

Fuel Reserve The amount of fuel beyond that required to fly to a destination that could be used in an emergency. The FAA regulations require at least a 30 minute reserve during the day and 45 minutes at night for VFR flight.

Gallons per hour The rate at which fuel is removed from the tanks and burned by the engine in one hour's time. GPH is the usual means for calculating fuel consumption in light aircraft. Larger aircraft may use pounds of fuel per hour. Automobiles usually use mile per gallon.

Global Positioning System (GPS) The system that uses satellites in Earth orbit to locate a position.

Groundspeed The actual speed that an airplane travels across the ground after considering the effects of wind.

Hazardous Inflight Weather Advisory Service (HIWAS) The weather reporting system that issues Airmets, Sigmets, Center Weather Advisories, and Pireps.

Indicated Airspeed The speed of an aircraft that is indicated on an airspeed indicator.

Inertial Reference System (IRS) An internal navigation system that can determine position without outside radio reception. The system starts with a known position and then senses both direction of movement and acceleration. A complete IRS system would also have radio reception (VOR, DME, etc.) crosschecks to verify accuracy.

International Date Line The line that is both 180 degrees East and 180 degrees West Longitude.

Isogonic lines There is an angle between True and Magnetic North Poles at all locations on Earth. The Isogonic Line depicts the angle between the poles to be used for navigation calculations. The Agonic Line is located where the angle is zero degrees.

Lambert Conformal Conic Projection The method of illustrating the Earth on a chart using a cone shape. Sectional Charts use this method for best accuracy.

Latitude The lines of the Earth's coordinate system that run parallel to the equator. These lines are also called Parallels.

Long Range Navigation System (LORAN) A medium frequency navigation radio system that utilizes a chain of stations. A chain includes one master station and several secondary stations to determine position.

Longitude The lines of the Earth's coordinate system that run from True North to True South poles. These lines are also called Meridians.

Magnetic Course The True Course corrected for magnetic variation. The Magnetic Course is used when determining whether to fly even or odd thousand foot levels.

Magnetic North Pole The magnetic concentration of the Earth. The magnetic compass points to the Magnetic North Pole.

Mercator Projection The method of illustrating the Earth on a chart using a cylinder shape. Chart accuracy is only assured where the inside of the cylinder touches the round Earth. This leads to an ever increasing distortion the farther from the point of contact. Sectional Charts do not use this method.

Military Operations Area (MOA) Areas where military flight training takes place. No special authorization is required to fly through these areas but caution is advised.

Military Training Routes (MTRs) Cross-country courses that are routinely flown by military aircraft. No special authorization is required to fly through these areas but caution is advised.

Missing Wind problem The technique whereby a pilot can calculate the actual wind that the aircraft is flying through. The calculation requires that the pilot first determine the TC, TAS, GS, and WCA. See chapter 13.

Navigation Plotter The tool that is used to determine True Course and distance on an aeronautical chart. The plotter consists of a "protractor" part and a "straightedge" part.

Navigation Record The form that a pilot uses to organize navigation information so that it becomes accessible in flight.

Nondirectional Beacon (NDB) The low to medium frequency radio broadcast station that transmits a signal. The Automatic Direction Finder (ADF) in the aircraft receives this signal for use in navigation.

Omni Bearing Selector The compass face of a VOR radio that allows the pilot to select a particular radial.

Pilotage Navigation by landmarks on the ground (looking-out-the-window navigation).

Pressure Altitude The altitude read on an Altimeter when 29.92 inches of mercury is set in the Kollsman window. Pressure Altitude is used in various flight performance calculations.

Prime Meridian The starting point of the coordinate system: Zero degrees. This line runs through the Royal Observatory in Greenwich, England.

Prohibited Area An area where civilian flight is never allowed.

Radial The lines that "radiate" out from a VOR station in every direction. The VOR instrument in the aircraft can determine which of the possible 360 different radials the aircraft is on.

Radio Navigation The form of navigation using various radio transmissions to determine position. VOR, NDB, LORAN, and GPS are all examples of radio navigation.

Remote Communications Outlet An antenna that can be used to communicate with a distant Flight Service Station.

Restricted Area An area where civilian flight is limited. A pilot should contact the restricted areas controlling agency prior to entry.

Ring Laser Gyro The device that determines aircraft movement and acceleration in order to determine position without outside radio reception. See chapter 22.

Satellite Constellation The 24 satellites in Earth orbit that together provide world-wide GPS coverage.

Sectional Chart The chart that depicts elevations and aeronautical information with a scale of 1:500,000.

See and Avoid Concept The technique of aircraft collision avoidance where pilots see each other in flight and then maintain visual separation. This concept requires good visibility and good scan technique by the pilot.

Sigmet A weather advisory that is significant to all aircraft regardless of size. Severe icing, severe turbulence, and thunderstorms will trigger a Sigmet.

Special VFR The clearance, requested by a pilot, that allows the aircraft to be flown into controlled airspace even though the proper visibility and cloud ceiling may not be present. See chapter 25.

Top of Climb The highest an aircraft will climb during a cross-country flight. This is the point where the pilot switches over from climb performance to cruise performance information.

True Airspeed The speed that an aircraft moves through the air taking into consideration pressure altitude and temperature.

True Course The direction of a planned flight in degrees relative to True North.

True Heading The direction of a planed flight in degrees relative to True North and after applying any necessary wind correction angle. The difference between a Course and a Heading is the wind drift correction.

True North Pole The geographic point around which the Earth revolves on its axis.

Warning Area The area over international waters where hazards to civilian flight exist from military operations such as controlled firing and bomb run practicing.

Wind Correction Angle The "crab" angle that pilots use to compensate for wind drift.

Wind Face The side of a manual flight computer that is used to calculate groundspeed, wind correction angle, and True Heading.

Wind Triangle The "pencil and paper" method of determining groundspeed, wind correction angle, and True Heading.

Winds and Temperatures Aloft Forecast The weather forecast that predicts the wind speed, wind direction, and air temperature at various levels above the surface. This forecast is used in flight planning to predict groundspeed and wind correction angles.

Variation Since the Magnetic and True North Poles are not in the same location, two lines can be drawn from any position on Earth to each of these poles. The angle between the two lines is the Variation. The addition or subtraction of Variation from a True Heading produces a Magnetic Heading.

Very High Frequency Omni Directional Range (VOR) The radio station that transmits a Reference Phase and a Variable Phase signal so that VOR radio receivers in flight can determine location relative to the broadcast station. See chapter 15.

VFR Flight Plan The form that describes features of a proposed flight. The FAA uses the Flight Plan to ensure the safe arrival of VFR flights.

VFR Terminal Area Chart An aeronautical chart of a high-density traffic area. This chart depicts the surface and features larger than a Sectional Chart for easier location identification. The scale of a VFR Terminal Area Chart is 1:250,000.

VOR Cross Check The procedure of using two VOR stations to locate an aircraft's position. The aircraft can be on a radial of the first VOR station and at the same time be on a radial from the second VOR. Where these radials cross is the location of the aircraft.

VOR Test Facility (VOT) A VOR broadcast station that is designed to test the accuracy of an aircraft's VOR radio receiver.

World Aeronautical Chart The aeronautical chart that is used for long trip planning. The scale is 1:1,000,000.

Index

W

ABOUT THE AUTHOR

Paul A. Craig, an experienced pilot, is the Chief Flight Instructor at Middle Tennessee State University in Murfreesboro, Tennessee. He holds eleven FAA Flight Certificates, including Airline Transport Pilot, and a Gold Seal Instrument and Multiengine Flight Instructor Certificate. He is completing a Doctor of Education Degree and holds the Master of Aerospace Education Degree. Craig is also the author of *Multiengine Flying, Second Edition*; *Be a Better Pilot*; and *Stalls and Spins* (McGraw-Hill).